Pipes, Mains, Cables and Sewers

Pipes, Mains, Cables and Sewers

Fifth Edition

H W Wilkinson, LLM, PhD

LONGMAN

ISBN 0 85121 602 1

First Edition 1967
Fifth Edition 1989

Published by
Longman Law, Tax and Finance
Longman Group UK Ltd
21–27 Lamb's Conduit Street
London WC1N 3NJ

Associated Offices
Australia, Hong Kong, Malaysia, Singapore, USA

A CIP catalogue record for this book is available from the British Library

Printed and bound in Great Britain by
Biddles Ltd, Guildford and King's Lynn

Contents

Preface

The law is stated as at 5 April 1989 but in the chapters on Electricity, Sewers and Drains, Water and Land Drainage the position both before and after privatisation is given.

It is assumed that the Electricity Bill and the Water Bill of 1988/89 have been enacted in the general form set out in the Bills.

H W Wilkinson
April, 1989

Table of Cases

Table of Statutes

Table of Statutory Instruments

Part I

Laying by Private Agreement, Implication or Prescription

Chapter 1

The Necessary Characteristics of a Pipe Easement

In *Re Ellenborough Park* [1956] Ch 131, at 163, the Court of Appeal accepted that the essential requirements of a valid easement at common law are set out in *Cheshire's Modern Law of Real Property* (now *Cheshire and Burn's Modern Law of Real Property*, 14th ed (Butterworths 1988), pp 492–6). These requirements are:

(*a*) There must be a dominant and a servient tenement.
(*b*) The easement must accommodate the dominant tenement.
(*c*) The dominant and servient owners must be different persons.
(*d*) The right must be capable of forming the subject-matter of a grant.

The characteristics are fully treated in *Gale on Easements*, 15th ed (Sweet & Maxwell 1986), pp 6–30 and P. Jackson, *The Law of Easements and Profits* (Butterworths 1978), pp 5–21. Only those aspects which particularly concern pipes will be discussed here. It will be convenient to follow the order of classification already given.

1 Dominant and servient tenements

To constitute an easement in private law there must be a dominant tenement to which the right belongs and a servient tenement upon which the obligation is imposed (*Mounsey* v *Ismay* (1865) 3 H & C 486, per Martin B, at 497). There can be no easement in gross (see (1980) 96 LQR 557, M F Sturley, criticising the weak theoretical basis of the rule).

Where there is an express grant the two tenements must be defined by the deed creating the easement, or be capable of definition through evidence of the circumstances surrounding the transaction: *Johnstone* v *Holdway* [1963] 1 QB 601 (CA) (identification of the dominant tenement by evidence of material facts at the date of execution when the deed failed to identify); *The Shannon Ltd* v *Venner Ltd* [1965] Ch 682 (CA) (evidence of the intentions of the purchaser when the deed was made was admitted to help to construe words used in the deed); *Bracewell* v *Appleby* [1975] Ch 408 (the fact that a right of way was over a private road serving private houses was relevant in construing a grant over the road of 'a right of way of the fullest description').

Where the grant is by prescription or implication, evidence of the extent of the user enjoyed over the statutory period or of the circumstances of the implication will delineate the tenements. A person claiming a pipe easement will usually be able to point out what land is burdened by tracing the route of the pipe, and the land benefited should be apparent from the place where the pipe terminates on his land and the function it fulfils.

The dominant tenement is usually corporeal property, a plot of land or a building served by a cable or pipe. In *Bass* v *Gregory* (1890) 25 QBD 481, the dominant tenement was an underground cellar used for brewing and the owner was held to have an easement for a current of air to come along to it through a ventilation tunnel under the servient tenement.

There is some authority for saying that an easement, which is incorporeal, can be appendant to incorporeal property — a right attached to a right. In *Hanbury* v *Jenkins* [1901] 2 Ch 401, a person claimed a right to a fishery over part of the River Usk and a right of way along both banks for angling purposes. He established his incorporeal right to a fishery, the servient tenements being the banks over which the right of way ran. *AG* v *Emerson* [1891] AC 649 was quoted by Buckley J (at 411), as establishing that the owner of the fishery is presumed, in the absence of contrary evidence, to own the soil of the river bed and this is the dominant tenement, but he added that, even if this were not so and the right to the fishery were incorporeal only, the incorporeal right of way could be appendant to it if there was agreement between the rights 'in nature and quality so as to be capable of union without any incongruity' (at 422). This reasoning could be used to claim for the owner of a pipe the right to go onto the land over which it runs to inspect and repair it, but the objection that the presumption concerning the ownership of the bed of a river does not apply to the soil supporting a pipe is a cogent one. A sounder argument to justify such an entry on land is that the grant of an easement carries with it all ancillary easements necessary for its exercise and enjoyment (*Bulstrode* v *Lambert* [1953] 1 WLR 1064), and this would include reasonable inspection and repair.

1.1 Dominant tenement

The dominant tenement may comprise both corporeal hereditaments (land and buildings) and incorporeal hereditaments (rights over land). In *Re Salvin's Indenture* [1938] 2 All ER 498 a water company had a right by deed to lay water pipes through a person's land. The question arose whether this was an easement or merely a personal licence and therefore unenforceable against any but the parties to the original deed. The water company owned no land contiguous to the burdened land, so the landowner contended that this was a claim to an easement in gross, as there was no dominant tenement. Farwell J held that there was a valid easement, the dominant tenement being the water undertaking itself, consisting of corporeal hereditaments (lands owned by the company for the purpose of its objects) and incorporeal hereditaments (the rights acquired in the lands of others, such as to lay pipes).

In view of this decision the practitioner may feel justified in advising a client that the combination of a landholding connected in some way, however remote, by pipes to the servient land and a right granted to lay his pipes there will constitute a dominant tenement sufficient to support a claim to an easement.

The unfortunate aspect of the case is that it is difficult to imagine any circumstances in which it would be held that there was no grant of easement in a similar situation unless the parties expressly agreed and stipulated against a grant. Possibly the judge was impressed by the consequences of holding that the water company's widespread system of pipes was laid under personal licence only. *Hanbury* v *Jenkins* [1901] 2 Ch 401 does not seem to have been cited in *Re Salvin*. It is suggested that the doctrine of non-derogation from grant could have been properly used in *Re Salvin* to prevent the successor in title to the grantor from obstructing the water company's rights (*Cable* v *Bryant* [1908] 1 Ch 259, per Neville J, at 264).

1.2 Servient tenement

The servient tenement will always be corporeal property, a plot of land or a building burdened in some way by the right of the dominant owner. In many cases, although it is clear which is the servient land, it is not clear how much of it is to be regarded as burdened. For example, if the servient land is a five-acre field over one strip of which a pipe runs by virtue of an easement, can the dominant owner claim that because he has an easement the servient owner could not sell off, say, four acres over which the pipe does not run, free of all burden? Either the vendor who sells off land should stipulate that the part is sold subject to the easement 'if and so far as' it affects the plot (which is rather vague conveyancing) or the purchaser should satisfy himself that it does not affect the plot and take a conveyance free from it, if necessary by getting the dominant owner to join in and release any possible right over the part sold off.

In most cases the easement will only burden materially that part of the land over which the pipe runs to a sufficient width and extent to ensure access to it to the dominant owner for the purpose of inspection and repair and to a sufficient depth and width to afford support to the pipe (*Newcastle-under-Lyme Corporation* v *Wolstanton Ltd* [1947] Ch 427). So the dominant owner can restrain a use of the servient land which will prevent him from enjoying his easement or make enjoyment substantially more difficult than previously. In *Goodhart* v *Hyett* (1884) 25 Ch D 182, the court restrained the servient owner from building over the line of a pipe which served the dominant land because the exercise of the right of the dominant owner to inspect and repair the pipe after the building was erected would have involved him in substantially more work and expense than if there were no building there.

The whole of the five-acre field mentioned in the example is burdened in the sense that the existence of the easement upon it may prevent its being sold as a complete unit as readily as if there were no easement. A plot of land subject to a right of way cannot be as easily sold as a site for a factory or department store as if the right did not exist. The Law Reform Committee (14th Report, Cmnd 3100 (1966)) has suggested that the Lands Tribunal should have power to discharge an easement or substitute one easement for another (para 97) but no legislation has resulted.

2 Easement must accommodate dominant tenement

In *Re Ellenborough Park* [1956] Ch 131, per Evershed MR, at 170, it was said that a right claimed over the land of another cannot be an easement unless it accommodates and serves the dominant tenement and is reasonably necessary for the better enjoyment of it; the right must have a necessary connection with the dominant tenement. In *Hill* v *Tupper* (1863) 2 H & C 121, the plaintiff leased land adjoining a canal, with the exclusive rights of running pleasure boats on it. The defendant, who had an inn on the canal, also let out boats on it. The plaintiff sued for interference with his 'easement'. He failed, the court saying that his claim was unconnected with his use and enjoyment of his demised land. Pipes will readily be seen to accommodate the premises they serve. The supply of water and the carrying away of effluent are reasonably necessary for the better enjoyment of premises. In *Wong* v *Beaumont Property Trust Ltd* [1965] 1 QB 173 (CA), a claim to affix a ventilation duct for kitchens onto adjoining premises was held to be an easement; in this case the duct was absolutely necessary for the enjoyment and continued use of the premises, under local authority requirements.

It is quite common to find older-type adjoining houses sharing chimneys and chimney stacks, with their flues leading into the same chimney. In this case there will usually be mutual easements of user and support and mutual obligations of repair. The mutual easements are reasonably necessary for the enjoyment of both premises (*Jones* v *Prichard* [1908] 1 Ch 630).

Cases could be envisaged where the right claimed does not accommodate the dominant tenement, for example, a claim to water one's cattle from a pipe on nearby private land. In *Re Ellenborough Park*, above, it was said that a right to a seat at Lord's cricket ground or to visit the zoo could not be an easement annexed to the ownership of a private dwelling-house, there being no sufficient nexus between the right and the enjoyment of the ownership (Evershed MR, at 174). It was held in *Rance* v *Elvin* (1983) 49 P & CR 65 that where water was supplied to plot *A* through a meter on plot *B* the owner of plot *A* could not claim an easement to have the water come to his land. The supply depended upon the owner of plot *B* being willing to continue to take the water and pay the charges. To impose a positive obligation on him to go on doing so would breach the principle that save in the anomalous case of an easement of fencing (*Lawrence* v *Jenkins* (1873) LR 8 QB 274) a right in the nature of an easement can only impose a negative obligation on the servient owner. Even though the claimed dominant owner was willing to pay his share of the charges for the water, they could not impose an easement where none was possible. It was further held on appeal in *Rance* v *Elvin* (1985) 50 P & CR 9 (CA) that though the owner of plot *B* had no obligation to ensure that water came through the pipes the owner of plot *A* had an easement for the passage of any water which did come through. For the distinction between a right to a supply of water and a right to the passage of water see *Schwann* v *Cotton* [1919] 2 Ch 459 (CA).

3 Dominant and servient owners

The dominant and servient owners must be different persons. Anyone who, for example, lays an electricity cable across his own land is not granting himself an easement but is exercising an incident of ownership *Metropolitan Railway Co* v *Fowler* [1892] 1 QB 165, per Lord Esher MR, at 171).

Can one person, however, in one capacity, acquire an easement for himself in another capacity, so that it continues when he disposes of the servient land? This question was raised, though not conclusively answered, in *Ecclesiastical Commissioners for England* v *Kino* (1880) 14 Ch D 213, where the question was whether windows of a church could acquire a right to light by prescription over the glebe lands adjoining. Both were vested in the rector, the church in him as trustee, the glebe lands in him as beneficial owner. An interlocutory injunction was granted to restrain building and all three members of the Court of Appeal indicated their view that the rector as beneficial owner could have burdened the glebe lands with a right to light for windows of the church of which he was trustee. But presumably in the converse situation, if the rector as trustee sought to burden the trust land for the benefit of the glebe land, he would be restrained from a dealing to his own advantage (*Ex parte Lacey* (1802) 6 Ves 625).

Other possible double relationships exist: sole personal representative against sole beneficiary, family company against husband and wife who are sole shareholders. Despite the obiter dicta in *Kino's* case, it is suggested that the practitioner cannot safely advise that a person can acquire easements for or against one or other of his legal personalities in this way. The rule in *Wheeldon* v *Burrows* (1879) 12 Ch D 31 should be borne in mind and if the 'double personality' owner sells off part of the land he should expressly reserve over the land sold any rights which he wishes to continue to enjoy as easements. A person without capacity cannot grant an easement expressly or impliedly. In the case of *Re St Clement's, Leigh-on-Sea* [1988] 1 WLR 720 it was held that the incumbent of a church had no power to grant impliedly an easement over the churchyard. For that a faculty of an ecclesiastical court was needed.

A tenant cannot acquire by prescription an easement over land of his landlord not demised to him, because the occupation of the tenant is in law the occupation of the landlord (*Gayford* v *Moffat* (1868) 4 Ch App 133, per Lord Cairns LC at 135). The exception is that a tenant can prescribe for an easement of light, Prescription Act 1832, s 3.

A limited company is a legal person (*Salomon* v *Salomon & Co Ltd* [1897] AC 22). Company mergers and amalgamations may affect easements; if *A* Co Ltd has a right to run a pipe over the adjoining land of *B* Co Ltd and the companies merge the easement will be extinguished by unity of seisin (*Whalley* v *Tompson* (1799) 1 Bos & P 371, per Eyre CJ, at 375). The easement will not revive for the benefit again of the amalgamated company if the land of the former *B* Co Ltd is sold off. It should be expressly reserved afresh for the benefit of the retained land (*Wheeldon* v *Burrows* above).

Where there is more than one dominant owner enjoying an easement over the servient land, one of them cannot sue the other for causing physical damage to the subject of the easement by excessive user unless the damage is such as substantially to interfere with and obstruct the plaintiff's own enjoyment of the easement (*Weston* v *Lawrence Weaver Ltd* [1961] 1 QB 402).

4 Right must be capable of forming subject-matter of grant

The following considerations are important in this connection:

(*a*) Are the rights expressed in terms too wide and vague?

(*b*) Would they amount to joint occupation with the servient owner or substantially deprive him of proprietorship or legal possession?

(*c*) Do they possess qualities of utility and benefit (*Re Ellenborough Park* [1956] Ch 131, per Evershed MR)?

Furthermore, just as the intending grantor must be capable of granting (*Mulliner* v *Midland Railway Co* (1879) 11 Ch D 611), so the proposed grantee must be capable of taking a grant. A grant cannot be made to a company with no power to hold easements (*National Guaranteed Manure Co* v *Donald* (1859) 4 H & N 8). Nor can there be a grant to an uncertain and fluctuating body such as the electorate of Birmingham, although in an old decision,

Gateward's Case (1607) 6 Co Rep 59 *b*, the inhabitants of a town established an 'easement' of way by custom over land. This would nowadays be called a public right of way and not an easement. A person with an interest in land cannot grant an easement over it in excess of his interest; he cannot grant an easement in fee simple over land on which he has a 99 year lease, and the grant will take effect for the residue of his term. If he subsequently acquires the reversion, this will not enlarge the easement (*Booth* v *Alcock* (1873) 8 Ch App 663). Similarly, a grantee cannot acquire an easement if it would be greater than his own interest in the land benefited (*Wheaton* v *Maple & Co* [1892] 3 Ch 48).

4.1 Vagueness

A claim will not be allowed if it cannot be defined. For example, a right of air to premises can be acquired over the land of another provided that the air flows in a defined channel, such as down a passage or tunnel (*Chastey* v *Ackland* [1895] 2 Ch 389), or provided that the air flows to definite windows or apertures on the dominant premises (*Hall* v *Lichfield Brewery Co* (1879–80) 49 LJ Ch 655). Only the former type of right concerns pipe easements as it constitutes in effect a right to have a 'pipe' (tunnel or channel) maintained on someone else's land. In *Chastey* v *Ackland*, above, Lopes J said (397) that a claim must be shown to have the air pass through some definite channel constructed for the purpose of containing and communicating it. It is suggested that this is too narrow a proposition and that a right could be acquired for air down a roofed-in passage between two buildings, or down a narrow passage not roofed in between tall buildings (sometimes called a 'ginnel' in Lancashire and a 'snickelway' in Yorkshire) even if they were not originally constructed for the purpose of communicating the air. There can, of course, be no claim to an easement for air which passes at large over another's land (*Webb* v *Bird* (1861) 10 CB (NS) 268). Such a claim is too vague and the servient owner would have no means of obstructing it and thus of preventing the acquisition of an easement (*Harris* v *De Pinna* (1886) 33 Ch D 238, per Bowen L J, at 262). But such a right could be claimed under the doctrine of non-derogation from grant, *Lyme Valley Squash Club Ltd* v *Newcastle-under-Lyme Borough Council* [1985] 2 All ER 405.

There can be an easement which is only enjoyed occasionally. From *Penn* v *Wilkins* (1974) 236 EG 203 it appears that a right could be recognised to lay temporary pipes through a neighbour's land to the dominant owner's cess-pit and to pump out the effluent to a tanker standing in the road, whenever the cess-pit needed emptying. The claim under s 62 of the Law of Property Act 1925 failed, because the 'right' was not enjoyed when the purchaser bought, the cess-pit being constructed after purchase.

If water percolates underground in an undefined channel no one can claim an easement to have it come to him through his neighbour's land; the claim is too vague. His neighbour could lawfully collect and use it, even if the effect were to stop the percolating supply, *Acton* v *Blundell* (1843) 12 M & W 324, or to cause the collapse of his land, *Stephens* v *Anglian Water Authority* [1987] 1 WLR 1381 (CA). The rule is otherwise if the water flows in a defined underground channel; here a claim against an abstracting neighbour will lie (*Chasemore* v *Richards* (1859) 7 HL Cas 376). This is a less usual type of pipe easement, not a claim to use, and send water along an underground course on someone else's land, but to have water come along the course to one's land and to prevent the neighbour from abstracting the flow or polluting it (*Stockport Waterworks Co* v *Potter* (1864) 3 H & C 300). It includes in many cases also the more usual right to pass the water on along someone else's land.

4.2 Joint occupation

A person who makes such a wide claim to use land that in effect he demands the full beneficial ownership will not succeed in establishing that he has an easement over it (*Copeland* v *Greenhalf* [1952] Ch 488). But temporary exclusion of the servient owner is a feature of many easements, for example, when a kitchen is shared (*Heywood* v *Mallalieu* (1883) 25 Ch D 357) or when there is shared use of a lavatory in an office (*Miller* v *Emcer Products Ltd* [1956] Ch 304). In contrast, in *Wright* v *Macadam* [1949] 2 KB 744, a right to use a coal shed, apparently virtually excluding the owner, was accepted as a right passing under Law of Property Act 1925, s 62. Unless the servient owner himself connects into and uses the pipe of the dominant owner on his land the dominant owner will have exclusive use of the part of the servient land occupied by his pipe. This will not bring him into conflict with *Copeland* v *Greenhalf* since he does not claim, in effect, beneficial ownership of the servient land as a whole, nor does he substantially deprive the servient owner of proprietorship or legal possession. In many cases the existence of a pipe through land will have no restrictive effect at all on the use to which it can be put;

crops can be grown over it if it is well buried and cattle can graze over it. If the pipe does interfere with the surface use, the question to what degree the servient owner is ousted from his enjoyment becomes important; relevant considerations are the amount of the exclusion (for example, the size of the pipe in relation to the area of the land), the permanence or otherwise of the exclusion and the length of time it continues or is likely to continue.

4.3 Utility and benefit

It would be unusual to find a case of a claim to a pipe easement which was not of utility and benefit to the dominant tenement. Even a claim to run a pipe over someone else's land to convey water to an ornamental fountain on one's own land would probably be upheld. The right to use a part as a pleasure garden was held in *Re Ellenborough Park* [1956] Ch 131 to be a right of utility and benefit and a beneficial attribute of residence in an adjoining dwelling-house. In this case some doubt was expressed by the court (177–9) whether the alleged distinction between a right of utility and benefit on the one hand and 'mere recreation and amusement' has any justifiable basis in law (it has its origin partly in the case of *Mounsey* v *Ismay* (1865) 3 H & C 486). The categorisation must be a false one, for recreation is a benefit and amusement is of utility. This requirement is probably inoperative in the sense that the alleged easement will not be held invalid under this head alone; if it fails it will do so under one of the other heads also – for vagueness, for joint ownership or for being ultra vires, for example.

It was stated by the court in *Re Ellenborough Park*, above, that the categories of easement are not closed and in that case a new easement was recognised but because the law has developed fully over many years it is unlikely that many more will be found. In *Phipps* v *Pears* [1965] 1 QB 76 (CA) it was held that there was no easement for protection from the weather by one building for another, followed in *Marchant* v *Capital and Counties Property Co Ltd* (1982) 263 EG 661 (but reversed on other grounds on the interpretation of an award under the London Building (Amendment) Act 1939 at (1983) 267 EG 843 (CA). There may be a benefit for protection from the weather derived incidentally if one building has an easement of support from another (*Bradburn* v *Lindsay* [1983] 2 All ER 408).

Chapter 2

The Nature of the Right to Lay and Use Pipes

1 As an easement

Some conflict of view has existed on whether the right to lay a pipe in another's land is an easement or not.

The right to the exclusive use of a water pipe running through adjoining land was recognised as an easement in *Goodhart* v *Hyett* (1884) 25 Ch D 182, per North J, at 186; and in *Nuttall* v *Bracewell* (1866) LR 2 Ex 1, per Martin B, at 10, it was said that the right to a flow of water in a goit is a well-known easement.

On the other hand it was held in *Bevan* v *London Portland Cement Co Ltd* (1892) 67 LT 615 that a person who had constructed and enjoyed the exclusive use of a tunnel under land adjoining his own land had gained a good title to the tunnel itself by limitation and that the enjoyment was not a mere easement. And in *Metropolitan Railway Co* v *Fowler* [1893] AC 416 it was held that for rating purposes the right to make a tunnel was the grant of a hereditament and not an easement. A case similarly decided concerning the right of drainage through a watercourse is *Holywell Union* v *Halkyn Drainage Co* [1895] AC 117. But the issue here was not between dominant and servient owners but between the taxpayer and the Revenue.

The majority of easements are of two kinds; they are either a right to do something on someone else's land, such as to go across it on a defined track (*Cannon* v *Villars* (1878) 8 Ch D 415), or a right to prevent him from doing something on his land which will affect his neighbour's land, such as building so as to restrict light (*Colls* v *Home & Colonial Stores Ltd* [1904] AC 179) or excavating so as to remove support (*Dalton* v *Angus* (1880–1) 6 App Cas 740). In these cases, although the servient owner is limited in his use of his land, he is not excluded from any of his land. The pipe easement is different. It consists of the right to put something in or on someone else's land, to maintain it there and to use it there, for example, by passing liquids along it. The right to tunnel involves in addition removing part of the neighbour's land and passing through the cavity thus made. These pipe easements in effect exclude the owner from the part of his land which is occupied by the pipe or tunnel. Now as a general rule a grant of the exclusive use of a piece of land takes effect as a grant of the land itself for all the estate therein of the grantor (*Reilly* v *Booth* (1890) 44 Ch D 12). The courts will not uphold a claim to an alleged easement which in effect amounts to a claim to exercise the rights of full ownership (*Copeland* v *Greenhalf* [1952] Ch 488). But in view of the other rights already recognised as easements which involve exclusion of the owner permanently from part of his land, for example a right to construct a yard on another's property and store goods there (*Mast* v *Goodson* (1772) 3 Wils 348), and a right to project a timber staging over adjoining land (*Harris* v *De Pinna* (1886) 33 Ch D 238), the general rule cannot be an absolute one. There is, as has been indicated, weighty judical authority for saying that the right to lay pipes in land and use them exclusively is an easement. Support is provided by *South Eastern Railway Co* v *Associated Portland Cement Manufacturers* (1900) *Ltd* [1910] 1 Ch 12, at 27; and in *Dunn* v *Blackdown Properties Ltd* [1961] Ch 433 the matter was treated as self-evident.

A validly created easement, held for the equivalent of a term of years absolute or of a fee simple absolute in possession, is a legal interest (Law of Property Act 1925, s 1(2)(*a*)). In *Land Reclamation Co Ltd* v *Basildon District Council* [1979] 1 WLR 106 (*affirmed* at (1979) 38 P & CR 528 (CA) a local authority validly granted under seal a right of way for a term of just under seven years. An easement for a term other than these takes effect as an equitable interest (Law of Property Act 1925, s 1(3)). A legal easement is valid even against a bona fide purchaser for value of the servient land without notice of it. An equitable easement will bind such a purchaser only if protected by registration as a land charge, Class D (iii) (Land Charges Act 1972, s 2), and in the case of land with a registered title, as a minor interest (Land Registration Act 1925, ss 3(*xv*), 49(1)). For an example of a right in the nature of an equitable easement, but arising out of mutual benefit and burden and not registrable as a land charge, see *E R Ives Investments Ltd* v *High* [1967] 2 QB 379 (CA).

Under the Law of Property Act 1969 the statutory period of a good root of title in contracts made after the Act takes effect (1 January 1970) is 15 years (s 23). The purchaser of land may thus not have actual knowledge of a land charge (such as an equitable easement) registered after 1925 but before his root of title commences. He is none the less bound by the land charge but if he suffers loss by reason of its existence he is entitled to compensation from the Chief Land Registrar (s 25). If he had actual knowledge he can have no compensation, but by s 24(1) registration of the land charge under the Land Charges Act 1925 or the Land Charges Act 1972 (1972 Act, s 18(5)), is not deemed to constitute actual knowledge, despite s 198 of the Law of Property Act 1925.

There can be different rights in different segments of the same land, as where a person has mining rights under the land of another. In *Williams* v *Usherwood* (1983) 45 P & CR 235 (CA) it was held that a person who lost his title to land by reason of the adverse possession of it by his neighbour nonetheless retained a right to go upon it to maintain his property and to drain his house through it. See also *Marshall* v *Taylor* [1895] 1 Ch 641 (CA).

The word 'easement' has been loosely used by the courts and Parliament. It has been applied to rights which may or may not have the characteristics of common law easements. For example, a landowner whose land was severed by a railway gained an easement to pass over it to reach his land on the other side, under the Railways Clauses Consolidation Act 1845, and this might well have the characteristics of a common-law easement. On the other hand, under s 7 of the Coastal Flooding (Emergency Provisions) Act 1953, a person may have a right to use a private road or 'formed way' to perform works of sea defence even if he owns no land to be benefited, and this will not have the characteristics. These 'statutory easements' will be so called in this book. The article of that name in (1956) *The Conveyancer*, 20, at 208, by Professor J F Garner, is very helpful. The precise use of terminology may help by reminding the practitioner that, in cases where rights over another's land are granted by private agreement (or implied or prescribed for), he must ensure that the easement is a valid one at common law, but that where the grant is by authority of statute the wording of the enabling Act is paramount. In *Mulliner* v *Midland Railway Co* (1879) 11 Ch D 611, the company, which had compulsorily acquired land under an Act, was held on the construction of the Act to have no right under it to go on to grant an easement over the land. In *Central Electricity Generating Board* v *Jennaway* [1959] 3 All ER 409, it was held that where the Electricity (Supply) Act 1919, s 22, granted the Board a right to pass overhead wires over land it conferred also by implication the right to the ancillary easement of erecting and maintaining on the land pylons to support them.

2 Under a licence

2.1 Types of licence

If a licence to enter on someone else's land is granted without payment or other consideration (a bare licence), the grantor can revoke it at will on reasonable notice without being liable in damages. In *Armstrong* v *Sheppard & Short Ltd* [1959] 2 QB 384 a person allowed another to lay a drain in his land and later gave about five weeks' notice to the grantee of revocation of the permission. This notice was held sufficient.

If a licence is granted for money or money's worth one party cannot terminate it otherwise than in accordance with the contract, provided that the other party is behaving properly and in accordance with the agreement (*Hurst* v *Picture Theatres Ltd* [1915] 1 KB 1). One might have a case where a market gardener pays for a licence from a neighbouring landowner to run temporary water pipes over the neighbour's land to serve some outlying fields during the season from 1 April to 15 September. If on 1 July the neighbour revokes the licence, so that the crops in the fields are ruined, the market gardener can claim damages for breach of contract. The *Encyclopaedia of Forms and Precedents* (the *Encyclopaedia*), 5th ed (Butterworths) Vol 12, 1269–70, and *Precedents for the Conveyancer* (Sweet & Maxwell), 1, 3034 give precedents.

A licence may be coupled with a proprietary interest in the land of the licensor, for example a licence to go on the land to cut grass and take it away. Rights to use pipes would rarely involve taking anything away from the land. If they did, such as a right to lay a pipe and extract brine through it, they would be profits à prendre and should be created by deed or prescription to be enforceable against a third party (Law of Property Act 1925, ss 1(2), 52(1)). If the interest is equitable it must be protected as a land charge under s 2(4) of the Land Charges Act 1972, or by notice in the case of registered land, if it is to bind a third party.

If a licensor has led the licensee to change his position (such as by spending money on his premises) in the expectation that the licence will be permanent, the licensor will be estopped from revoking the licence (*Dillwyn* v *Llewelyn* (1862) 4 De GF & J 517; *Inwards* v *Baker* [1965] 2 QB 29). In *Ward* v *Kirkland* [1967] Ch 194 a verbal permission to lay and use drains was held in the circumstances to be permanent and unlimited in time, as it would have been unconscionable for the grantor to revoke his permission when expenditure had been made in reliance on it, but in *Armstrong* v *Sheppard & Short Ltd* [1959] 2 QB 384 a verbal right to use a drain was held revocable on reasonable notice, although the grantor could not require the grantee to remove the drain which he had allowed to be laid in his land. If the adjoining landowners have granted each other mutual benefits, such as a right of way, of drainage or of allowing the encroachment of foundations, one of them will not be allowed to enjoy the benefit of the arrangement whilst rejecting the burden. He may not withdraw the licence unilaterally: *Hopgood* v *Brown* [1955] 1 WLR 213 (CA) (mutual drainage rights); *E R Ives Investments Ltd* v *High* [1967] 2 QB 379 (CA) (right of way and use of foundations).

2.2 Position of third parties

If the licensor sells his land, can the licensee enforce the licence against the purchaser? The grant of a licence is in general a transaction personal to the parties so that even a person who takes the land with notice of the licence is not bound by it. In *King* v *David Allen & Sons, Billposting Ltd* [1916] 2 AC 54 a licence to post bills on a cinema wall was held unenforceable by the licensee against the successor in title of the licensor, who had taken the premises with notice of the licence: *Midland Bank Ltd* v *Farmpride Hatcheries Ltd* (1980) 260 EG 493 (CA) should be contrasted.

Equity has intervened in certain situations to make licences enforceable by and against successors in title of licensors and licensees. Some of the cases have been mentioned above. For example, where there is an expenditure of money in the expectation (induced by the licensor) that the licence will be permanent, the successor in title with notice is bound by it (*Inwards* v *Baker*, above); similarly where there are mutual licences granted by adjoining landowners (*ER Ives Investments Ltd* v *High*, above).

Equity may regard the purchaser who takes with notice of a licence as being subject to a constructive trust. In *Binions* v *Evans*

[1972] Ch 359 (CA) a purchaser of premises bought them expressly subject to the right of an elderly lady to live in them for her lifetime rent-free. The purchaser was held not to be entitled to revoke her right and to turn her out, though he had made no contract or agreement with her. He was a constructive trustee of the premises to the extent of her interest in them.

It was the view of Lord Denning in *Bendall* v *McWhirter* [1952] 2 QB 466 (and in *Binions* v *Evans*, above) that even against a purchaser for value a contractual licensee who is in actual occupation of land by virtue of a licence has an interest which is valid, if not at law then in equity, against the successors in title of the licensor so long as the conditions of the licence are observed. It is not a legal interest in the land but a clog or fetter like a lien. The occupation by the licensee is sufficient notice of his rights to a purchaser (Law of Property Act 1925, s 14; Land Registration Act 1925, s 70(1)(*g*)). The doctrine has not found favour with the House of Lords (*National Provincial Bank Ltd* v *Ainsworth* [1965] AC 1175, at 1239, 1251). In contrast, Megarry and Wade, *The Law of Real Property*, 5th ed (Stevens 1984) p 808), changing its opinion from the 4th ed (1975) p 783, says 'all the indications now are that contractual licences are capable of binding successors in title as equitable interests (or perhaps as mere equities)'. Cheshire and Burn's *Modern Real Property*, 14th ed (Butterworths 1988) p 562, says after a review of the cases (and citing in support dicta in *Ashburn Anstalt* v *Arnold* (1987) 2 EGLR 71 (CA) that they 'provide no support for the view that contractual licences generally are binding on third parties'. However *Emmet on Title*, 19th ed (Longman), para 5.156 says 'It now appears that a person having a licence to use land may hold an equitable interest enforceable against a purchaser other than one who takes a legal estate for value without notice.' It was held in *Midland Bank Ltd* v *Farmpride Hatcheries Ltd* (1980) 260 EG 493 (CA) that a licence can bind third parties (there a mortgagee) who have merely constructive notice and not actual notice.

In *Cook* v *Minion* (1978) 37 P & CR 58 *A*'s predecessor in title permitted *B*'s predecessor in title to re-lay his drains so as to run through *A*'s drain. *A* was held bound to continue to allow *B* to do the same, since *B*'s predecessor had incurred expense in reliance on the permission.

Is a licence registrable under the Land Charges Act 1972? Section 2 of the Act authorises registration under Class D (iii) as an equitable easement of any easement, right or privilege created or arising after 1925 and being merely an equitable interest. In *E R Ives Investments Ltd* v *High*, above, the Court of Appeal seems to have held that a mutual licence (of way for the one party and of support of foundations for the other) was not registrable, and in *Shiloh Spinners Ltd* v *Harding* [1973] AC 691 the House of Lords held that an equitable right of re-entry on breach of covenant was not registrable. Perhaps Class D (iii) is confined to equitable easements and rights in the nature of easements and profits. If a licence has the support of a dominant tenement to be benefited by it, it is suggested that it is registrable under Class D (iii). Where the title to the land is a registered one, then if the licensee has drainage rights or water rights or is in possession he may be held to have an overriding interest which is not capable of entry on the register but subject to which all dispositions will take effect (Land Registration Act 1925, s 70(1)). Section 70(1) gives (amongst others) as overriding interests 'drainage rights, . . . watercourses, rights of water, and other easements not being equitable easements required to be protected by notice on the register' (para (a)) and 'the rights of every person in actual occupation of the land . . . , save where enquiry is made of such person and the rights are not disclosed' (para (g)). If he is not in possession and does not possess water rights he will not have an overriding interest. Any other type of licence over registered land seems not to be an overriding interest, and it is not clear whether it is registrable against the land as a minor interest under ss 3(*xv*) and 49(1) of the Land Registration Act 1925. Ruoff and Roper on *The Law and Practice of Registered Conveyancing*, 4th ed (Stevens 1986) pp 135–6) say that in unregistered land licences arising by estoppel are not registrable as land charges, but in registered land they can be protected by notice, restriction or caution, Land Registration Act 1925, ss 49(1) (f) and (2), 53 and 54.

Can the licence be assigned to a person who buys the licensee's land? A bare licence is personal to the grantee and will cease on an assignment (*Terunnanse* v *Terunnanse* [1968] AC 1086). A licence protected by equity can be enjoyed and enforced by the successor in title of the licensee (*Hopgood* v *Brown* [1955] 1 WLR 213). A licence coupled with an interest in land is assignable (*Muskett* v *Hill* (1839) 5 Bing NC 694, at 707–8 (licence to mine for and remove tin ore)). A contractual licence is not assignable unless the contract so provides (*Clore* v *Theatrical Properties Ltd* [1936] 3 All ER 483).

The case of *Re Salvin's Indenture* [1938] 2 All ER 498 (discussed in the previous chapter) is a very important one. If ever it is held to have been wrongly decided, many pipes now thought to be laid by virtue of easements may be found to be laid by virtue of personal licences. Problems of enforcement by and against successors in title would then become acute.

3 Under the doctrine of non-derogation from grant

The doctrine that a grantor may not derogate from his grant has been applied in certain cases to enforce obligations which could not ordinarily be protected as easements. A person entitled to an easement of light cannot object to any and every diminution made in his light by his neighbour, but in *Herz* v *Union Bank of London* (1854) 2 Giff 686, a person who took a lease of premises for the *express* purpose of carrying on his business of a diamond merchant and who needed a greater amount of light to his windows than most businesses would need was able to restrain his lessor from diminishing the light which he enjoyed and required. An example concerning pipes (in this case, a ventilation duct) is provided by *Wong* v *Beaumont Property Trust Ltd* [1965] 1 QB 173 (CA), where an assignee of a lessee restaurant-owner was held to be entitled to affix a duct from his kitchens to a wall of the assignee of the landlord on premises not included in the lease to comply with the requirements of the public health authority. The court found here an easement of necessity, but the grounds of the decision have been criticised by the writer (inter alios), who has suggested that the doctrine of non-derogation is a more appropriate ground ((1964) *Modern Law Review*, 27, 720). It is appreciated, however, that this suggestion involves an extension of the doctrine: from restraining a person who has made a grant from doing something more which will derogate from it, to not permitting him to derogate by *failing* to do something – eg to give his permission for a ventilation duct to be affixed to his wall. It is suggested that derogation can just as much occur by not doing, as by doing.

The doctrine can be applied against an assignee of the grantor's interest whether or not he is aware of the obligation (*Cable* v *Bryant* [1908] 1 Ch 259, per Neville J, at 264). So if a landowner sells off part of his land for the known purpose that a house can be built on the part sold, and the house is built with drains lawfully running through the part retained and the landowner sells the land retained to a person who commences excavations which will damage the drains, this person can be restrained under the doctrine from so doing.

Can an assignee of the grantee's interest invoke the doctrine against the grantor or the grantor's successor in title? This point is open. *Wong's* case, above, supports the view that he can. Further, since it has been decided that the obligation binding the grantor is enforceable against his successors (*Cable* v *Bryant*, above), then the successors of the grantee should acquire his rights so long as they use the premises for the same purpose as gave rise to the original benefit in the grantee. *Gale on Easements*, 15th ed, p 103, supports the view, but gives *Harmer's* case, below, as an authority, which does not support it. *Cook* v *Minion* (1978) 37 P & CR 58 now supports it.

The doctrine is not limited to physical interference with the enjoyment of the land demised; non-physical interference is included. Where a lessee took land for an explosive store and adjoining land was leased by the same lessor to a person who built so near to the store that the licence to store explosives was revoked by the Home Office, it was held that by the subsequent leasing the lessor had derogated from his first grant (*Harmer* v *Jumbil (Nigeria) Tin Areas Ltd* [1921] 1 Ch 200). This decision could be of importance where the first grant is of a right to lay a pipe carrying a dangerous or corrosive substance and a subsequent grant is made which may turn this into a more hazardous matter than previously, such as a grant of a plot of land on which fuel tanks are to be erected.

It will be unusual to find pipes laid affected by the doctrine or to find them laid by licence. They are far more likely to be laid by virtue of an easement. Accordingly, subsequent discussion will largely concern itself with pipe easements, though the observations may be relevant to the other types of case.

The doctrine of non-derogation from grant is thoroughly discussed in (1964) *Law Quarterly Review,* 80, 244 (D W Elliott).

4 Trespass

A person who connects his drain to someone else's without permission and uses it is liable for trespass even if the other suffers no damage. However, an injunction will not necessarily be granted to restrain the trespass. In *Cook* v *Minion* (1978) 37 P & CR 58 a person without permission connected the drainage of his house to a common pipe which served several houses. An injunction was refused since it would be oppressive to the defendant if he were ordered to alter the drainage arrangements of his house. Damages were awarded instead.

5 Common law rights of drainage

Where land is on the bank of a watercourse (whether the water flows on or below ground) the proprietor or occupier of the land is entitled to the flow and bound to accept it. He is not entitled to deprive landowners lower down of the flow, nor to pen it back on the land of his upstream neighbour, *Mason* v *Hill* (1833) 5 B & Ad 1, 110 ER 692.

Where land is not on the bank of a watercourse:

(a) where water does not run in a defined channel but runs freely over the surface or percolates through the soil a higher owner may appropriate the water (for example in a reservoir) no matter how long the lower owner has enjoyed it, *Broadbent* v *Ramsbotham* (1856) 11 Ex 602, 156 ER 971;

(b) the lower owner cannot complain of water coming from the higher land but is not bound to accept it and can put up barriers to it, even if this causes damage to the higher land, *Home Brewery Co Ltd* v *William Davis & Co (Leicester) Ltd* [1987] 2 WLR 117;

(c) the owner of land has no right to discharge, such as by pumping, water onto his neighbour's land which has come there by gravity, *Baird* v *Williamson* (1863) 15 CB (NS) 145, or which has come naturally but has been kept there artificially, *Whalley* v *Lancashire and Yorkshire Railway Co* (1884) 13 QBD 131, and is liable if he has collected it there and it escapes, *Rylands* v *Fletcher* (1868) LR 3 HL 330;

(d) an occupier may protect his land from floodwater, by sea or river, and build defences against it even if the result is that it flows with greater violence onto someone else's land but he may not injure his neighbour's land in getting rid of the water, *Whalley's case*, above.

Chapter 3

Acquisition of the Pipe Easement and Extent of the Right

1 Acquisition by deed or other express grant

A landowner's power to grant easements of whatever nature he chooses over his land is subject to certain restrictions.

First, planning permission will be required if the work to be carried out constitutes 'development'. The Town and Country Planning Act 1971, s 22(1), defines development as 'the carrying out of building, engineering, mining or other operations in, on, over or under land, or the making of any material change in the use of any buildings or other land'. Section 22(2) lists certain matters which do not constitute development, including the carrying out of works by a local authority or statutory undertaker for 'inspecting, repairing or renewing any sewers, mains, pipes, cables or other apparatus' (para (c)). This does not exempt such work by private persons and it does not include the laying of new services. By s 22(2) of the Act, the carrying out of works for alteration of a building by providing additional space therein below ground does constitute development, despite the general rule in s 22(2) that alterations to the interior which do not materially affect the external appearance are not development. Certain works of development may be 'permitted development' under the Town and Country Planning General Development Order 1988 which replaces the 1971 and 1981 Orders from 5 December 1988 (SI No 1813), in which case there is no need to obtain an express grant of planning permission before carrying them out unless the land is in an area of special control such as a National Park.

The permitted developments are given in Sched 2 and include:

(a) the enlargement, improvement or other alteration of a dwellinghouse up to stated sizes (Part 1);

(b) development on industrial land including the provision or

re-arrangement of a sewer, main, pipe, cable or other apparatus (Part 8, Class B);

(c) carrying out works for inspecting, repairing, or renewing any sewer, main, pipe, cable or other apparatus (Part 10, Class A);

(d) development authorised under any local or private Act for erecting or altering any bridge, aqueduct, pier or dam (Part 11, Class A);

(e) development by water authorities and for sewerage and sewage disposal (Parts 15 and 16);

(f) other parts authorise developments for providing or replacing sewers, drains or cables by water or hydraulic power undertakings (Part 17, Class C); gas suppliers (Part 17, Class F); electricity undertakings (Class G); tramway or road transport undertakings (Class H); mining operations (Part 19, Class A); British Coal mining development (Part 20); and telecommunications development (Parts 24 and 25).

A second restriction on the freedom to grant easements is that the landowner may not grant a right which is contrary to the public good, such as a right to drain into and pollute a stream contrary to public health regulations. In the case of water rights he cannot grant a right which will prejudice owners of other sections of the watercourse, unless they consent. Draining heated water into a river and so killing the fish is an example (*Pride of Derby & Derbyshire Angling Association Ltd* v *British Celanese Ltd* [1953] Ch 149).

A third requirement is the obvious one that the easement sought to be granted must have the characteristics of a common-law easement discussed in the previous chapter.

1.1 Creation of easement

It is essential to the grant of a legal easement that it is created by deed (Law of Property Act 1925, s 52(1)). If so made it is enforceable against all persons taking any interest in the servient land. It is not registrable under the Land Charges Act 1972. The only exception to the requirement for a deed is where the easement is granted or reserved under a disposition by will; a legal easement will arise when the assent is made by the personal representatives but will relate back to the date of death (Administration of Estates Act 1925, s 36(2), (4)). A grant made otherwise than by deed will take effect as an equitable easement which will depend on registration under the Land Charges Act 1972, s 2, under Class D (iii) for enforceability against persons acquiring a legal estate in the servient land for money or money's worth (s 4(6); Land Registration Act 1925, s 52(1)). Where the grantee has bound himself to buy land to be benefited by a grant of easement but has not at the date of the grant taken a conveyance, assignment or (in the case of registered title) transfer, it is suggested that the grant becomes completely effective and the legal easement arises when the conveyance or assignment is made, and in the case of a transfer when the transfer is registered at the Land Registry (Land Registration Act 1925, ss 20(1), 22(1)). Until then the grant takes effect in equity only (*Johnstone* v *Holdway* [1963] 1 QB 601 (CA)) but if the title is a registered one and the right is actually enjoyed it may take effect as an overriding interest, *Celsteel Ltd* v *Alton House Holdings Ltd* [1985] 2 All ER 562.

An order for specific performance may be made against a person who covenants to make sewers under his own land for the benefit of land which he has sold off for building and who fails to make them: in *Carpenters Estates Ltd* v *Davies* [1940] Ch 160 it was held that in such a situation, where the plaintiff could not reasonably build until the defendant had carried out his covenant to make the sewers which would serve the buildings, damages would not be an adequate remedy.

1.2 Construction of grant

It is a rule of construction of a grant of water easements, laid down by Jessel M R, in *Taylor* v *St Helens Corporation* (1877) 6 Ch D 264, at 270–1, that one must find out the meaning of the instrument according to the ordinary and proper rules of construction and that the maxim that in cases of doubt or obscurity the grant must be read against the grantor is of no force. In this case he pointed out that the grant of a watercourse, particularly when coupled with other words, might mean the easement for the running of water, the drain which carries the water or the land over which the water flows. He said that if the context did not show which was meant then the easement for the running of water was meant. Where water rights over land are created under a deed the nature and extent of the interests of the parties is regulated wholly by the deed, and the court cannot consider what rights they would have had as riparian owners if they had not made the deed (*Northam* v *Hurley* (1853) 1 E & B 665); the rule applies also to underground water in springs (*Whitehead* v *Parks* (1858) 2 H & N 870).

Where the meaning of the deed is clear, evidence cannot be given that it means something other than what it appears to mean, for that would be to contradict the deed. Only if the deed is not plain may the court consider the circumstances surrounding the making of the grant (*The Shannon Ltd* v *Venner Ltd* [1965] Ch 682, per Danckwerts L J, at 693). The court's task is to ascertain the parties' intentions by construing the deed, and it will consider (*a*) the circumstances surrounding the execution of the instrument, and (*b*) whether there have been words inserted limiting the type of user (*Cannon* v *Villars* (1878) 8 Ch D 415, per Jessel M R, at 420–1). In *Shayler* v *Woolf* [1946] Ch 320 (CA) the defendant conveyed a plot of land adjoining his own to Mrs Peacock for the erection of a bungalow and he covenanted to supply water to the premises from a pump on his own land and to keep the pump in good repair and working order. Mrs Peacock sold the bungalow to the plaintiff, Mr Woolf with the benefit of the covenant. The pump became blocked and a new one was made nearby but the defendant refused to continue the supply to the plaintiff, who then claimed specific performance. It was held that on the true construction of the agreement it was the intention of the parties that the benefit of the covenant should be assignable, otherwise it would cease on the death of the grantor, and the bungalow would at once be much reduced in value. The argument that the supply of water was one for personal services which equity would not enforce was rejected by the court, as was the argument that the duty to maintain the pump should not be enforced as it would involve the covenantor in renewing the pump completely from time to time. The court said that repair need not necessarily involve renewal.

In *St Edmundsbury and Ipswich Diocesan Board of Finance* v *Clark (No 2)* [1975] WLR 468 (CA) it was held that a reservation of a 'right of way' was ambiguous and must be construed by looking at the surrounding circumstances at the time of the grant. Here the path of the way appeared not to be a roadway but a derelict and overgrown path when the grant was made and therefore it was a way for passage on foot, not a vehicular way. Although a reservation was, by s 65(1) of the Law of Property Act 1925, technically a re-grant by the purchaser the question of whether there was any presumption either as against the vendor or the purchaser only fell to be considered when the court was unable on the material before it to reach a sure conclusion on the proper construction of the reservation. A purchaser can create an easement over the property which he is buying, by having it conveyed to him 'subject to' the easement which is to be created (1925 Act, s 65(2)); an unidentifiable person can take an interest in land even though he is not a party to the conveyance, s 56, 1925 Act, *Wiles* v *Banks* (1985) 50 P & Cr 80.

1.3 Drawing the grant

Identifying tenements. Although it is desirable to identify the dominant tenement which the easement will benefit, the grant will not be ineffective if this is not done, provided the court can satisfy itself by evidence of the material facts existing when the deed was made what was the whole object of the transaction (*Thorpe* v *Brumfitt* (1872–3) 8 Ch App 650). As a matter of prudent drafting, however, the dominant tenement should certainly be identified, as should the servient tenement (*Woodman* v *Pwllbach Colliery Co Ltd* [1915] AC 634). It is suggested that the same rules are applicable to ascertain the servient tenement as to ascertain the dominant tenement.

In *Land Reclamation Co Ltd* v *Basildon District Council* [1979] 1 WLR 106 (affirmed at (1979) 38 P & CR 528) where a right of way was granted for a term of years, by deed, Brightman J said that it was well established that a right of way could exist as a legal interest notwithstanding that the dominant tenement was not precisely identified in the deed of grant. 'The general purpose reference to the company's land in the lease is sufficient for this purpose, extrinsic evidence being admissible' (110).

It is proper conveyancing practice not only to identify the dominant land but also the owners or occupiers of it so as to make it doubly clear what land is benefited (Danckwerts L J in *The Shannon Ltd* v *Venner Ltd* [1965] Ch 682, at 692). *Kelly's Draftsman*, 15th ed, 323, contains a useful precedent.

The dominant and servient plots should be indicated on a plan unless they can safely and completely be identified otherwise, by street numbers for example. The line of the pipe should be indicated so that the places of its beginning and end are clear. For a declaration as to party walls and drains where part of land is sold off see *Kelly's Draftsman*, 15th ed, p 331.

If an easement is granted to serve dominant tenement *A*, can the same easement be used to accommodate dominant tenement *B* which the easement owner later acquires? The need for tight drafting is shown here. The common-law position was set out by Romer LJ in *Harris* v *Flower* (1904) 74 LJ Ch 127, where he said that if a right is granted for the enjoyment of plot *A* and the grantee owns or acquires plot *B* he cannot use the way to plot *A* to accommodate plot *B* also. The principle would apply where a right to lay a pipe to plot *A* was granted and plot *B* was sought to be connected onto the pipe. In *Bracewell* v *Appleby* [1975] Ch 408 a person was granted a right of way 'of the fullest description' over a private road to gain access to his plot (number 3) in a cul-de-sac. Six plot owners shared the private way to reach their plots in the cul-de-sac. The owner of plot 3 bought some land-locked land which was behind it (plot 2a) and made a way to plot 2a over his plot 3, and thence over the way in the cul-de-sac, despite protests from other plot owners. The additional use was held not to be permitted by the grant and damages were awarded. An injunction to forbid use was refused, as it would have made the house on plot 2a unusable. The correctness of the decision to refuse an injunction in *Bracewell* was strongly doubted by Scott J in *Anchor Brewhouse Developments Ltd* v *Berkley House (Docklands Developments) Ltd* [1987] 2 EGLR 173.

Purpose and extent. The purpose and extent of the grant should be stated. For example, in *Land Reclamation Co Ltd* v *Basildon District Council* [1979] 1 WLR 106 a right of way was validly granted by deed for a company between stated times to pass along a private road 'for all purposes in connection with the use of the Company's land for the dumping and disposal of waste materials'. If a pipe is to drain away rainwater the servient owner should stipulate this so that it is not subsequently adapted to be used for sewerage or other polluted water. If only a general description is given, such as 'for the purpose of drainage of the dominant premises', then the pipe can be used for any purpose which it will accommodate (*White* v *Grand Hotel Eastbourne Ltd* [1913] 1 Ch 113; *Keefe* v *Amor* [1965] 1 QB 334 (CA). But even though a grant is in wide terms a particular use could be forbidden if to allow it would restrict the use which could be made by others who also have a right to use the land: *Jelbert* v *Davis* [1968] 1 WLR 589 (CA) (grant of right to use a way 'in common with all others having the like right'; proposal to put 200 caravans on adjoining land, with their occupiers using the way; held: not permissible for these numbers to use the way, their use would interfere with those 'having the like right').

In *Rosling* v *Pinnegar* (1987) 54 P & CR 124 (CA) a person bought a large house which had a right of way to be used 'in common with the purchaser and all persons deriving under him'. He restored the house with a grant from public funds which required him to open it to the public on not less than 30 days in the year. The right of way was originally granted 'for all purposes'. It was held that the mere attraction of visitors along the way was not itself a breach of the terms of the easement but since on the facts user had been such as to interfere unreasonably with use by others who were entitled to the way an injunction would issue to control and limit the use.

If the servient owner wishes to limit the intensity of user he should stipulate this also; the dominant owner might build other premises on his land and seek to drain these through the same pipe, and the result might be overloading of the pipe with danger of fracture or overflow, as in *Wood* v *Saunders* (1875) 10 Ch App 582. Unless he stipulates that the pipe is to serve the particular dominant premises existing at the date of the grant, the servient owner may have difficulty in preventing increased intensity of user. Relevant questions are whether the grant was a limited one or a general one, and whether the use complained of goes outside the terms of the grant *Hurt* v *Bowmer* [1937] 1 All ER 797). In *Phillimore* v *Watford Rural District Council* [1913] 2 Ch 434 a person with a right to drain water over someone else's land was restrained from draining sewage over it.

Care should be taken to state precisely the extent of the grant. Where a right is to be granted to drain foul water into a sewer for example, it is of little use to grant a right to pollute 'to the extent and in the manner and with the materials at present being discharged' if it has not been established what this manner, extent and materials consist of. Some means should be found of making an agreed measure or description which should then be contained in the grant itself or incorporated into it by reference, for example, 20,000 gallons per day, or between the hours of 10 pm and 6 am. For a precedent which allows drains which may be made in the future (within the perpetuity period) to be connected, see *Precedents for the Conveyancer*, 2, 8099. In *Trailfinders* v *Razuki* (1988) 30 EG 59 it was held that a lease which reserved to the landlords the free passage of electric current from other buildings of the landlords through the wires and conduits 'which are now or hereafter during the term' may be in, over or under the demised premises, did not entitle the landlords to lay and use new cable for a computer link-up to their adjoining premises.

It is not possible to give fixed meanings to certain standard conveyancer's phrases; the meaning will differ with the situation. For example, in *Collins* v *Slade* (1874) 23 WR 199, a grant of a right of way 'as at present enjoyed' was held to mean that user was restricted to certain daylight hours. In *Hurt* v *Bowmer*, above, the phrase in a grant of a way over a field 'as at present enjoyed' was held to have no relevance to the purpose for which the way was used at the time of the grant but only to the quality of user – whether on foot or with vehicles. Since the plaintiff's complaint was over the quantity of user since the grant and not over its quality and type, which had remained the same, he was unsuccessful.

In *Polden* v *Bastard* (1865) LR 1 QB 156, a person owned two adjoining houses. She occupied one and let the other to a tenant, Thomas Answood, who used the water pump which was on her premises as he had no water supply. The owner devised her own house to the plaintiff and the other 'as now in the occupation of Thomas Answood' to her niece, who later conveyed it to the defendant. The plaintiff refused access to the pump on his property and the defendant took water by forcible entry. It was held that he was not entitled to water from the pump. The devise had merely identified the property being conveyed; it had not evinced an intention to pass an easement to the devisee of the tenanted property. In this case there was no easement of necessity as the tenant could have sunk a well on his premises or have used a nearby river.

Where the right is given to a class of persons (such as to *A*, his heirs, assignees, tenants and licensees) it is a question of construction of the grant whether the list is exhaustive or illustrative (*Hammond* v *Prentice Bros Ltd* [1920] 1 Ch 201).

The grant of a particular type of easement may by implication put an obligation on the grantor. In *Liverpool Corporation* v *Irwin* [1977] AC 239, at 256–7, Lord Wilberforce suggested that if a right to use a passage was granted and the natural light was absent or insufficient 'to the extent that the easement is useless without some artificial light being provided, the grant should carry with it an obligation to take reasonable care to maintain adequate light'. The suggestion is novel. The general rule is that the only easement which can impose a duty to expend money is the 'spurious easement' of fencing (*Lawrence* v *Jenkins* (1873) LR 8 QB 274).

P Jackson, *The Law of Easements and Profits*, at 65 considers that Lord Wilberforce's suggestion could be better put upon the basis of nonderogation from grant.

Contents of deed. Normal provisions in a deed of grant are arrangements for laying, inspecting, cleansing and repairing the pipes (a right of entry will be granted to the dominant owner for this purpose), and for paying for the cost of laying the pipes and legal and surveyor's fees in connection with drawing the deed and obtaining any necessary consents. The deed should stipulate that the plans and specifications of the pipes are to be approved by the grantor before construction commences and that the works should comply with the plans. There should be a right to inspect them at various stages before they are covered, and laying should not proceed beyond each stage until approved. Construction should comply with the Building Regulations 1985 (SI No 1065) and the current British Standard Code of Practice. It may be thought wise to have a sum to cover the cost of the works deposited in a bank account in the joint names of the solicitors for the grantor and the grantee and a right to the grantor to use the moneys to have the laying completed should the grantee fail to do so within a specified time after due notice to complete. This will cover the contingency of the grantee going bankrupt. A precedent for the grant of an easement to lay a pipe and to drain through it into a septic tank is contained in the *Encyclopaedia*, 5th ed, p 1269; for general grants and reservations of rights of drainage in registered land see Ruoff and West, *Land Registration Forms* 3rd ed, (Sweet and Maxwell, 1983) pp 75, 76, 130, 131, 141, 142.

Further matters to consider are: the bore of the pipe and the material of which it is made (so that it will not leak), the imposition of a duty to repair and liability for damage caused by leaking, the depth of the pipe, to make sure that crop growing on the servient land is not hampered; reinstatement of the subsoil and topsoil when the pipe has been laid; reinstatement of any land drains affected by the pipe; and removal of the pipe if it ceases to be used. It should be ensured that building and development on the servient land is not prevented by the pipe's existence (*Goodhart* v *Hyett* (1884) 25 Ch D 182), and if the pipeline is likely to prevent development full compensation should be obtained as part of the consideration for the grant. It may be possible to cover this point by saying that the parties will try to agree an alternative route if planning permission for development is later obtained and if the development would be frustrated by the presence of the pipe and providing that in default of agreement the value of the loss of development benefit shall be settled by arbitration.

Future easements. If the original grant gives the dominant owner a right to divert to a route which remains to be chosen he may be held to have a future easement and if so his grant should be restricted to the perpetuity period to be a valid one (*Dunn* v *Blackdown Properties Ltd* [1961] Ch 433). It might, however, be saved by the 'wait and see' provisions of the Perpetuities and Accumulations Act 1964, s 3, even if not expressed to be limited to the perpetuity period, provided that the grant took effect after the date when the Act came into force on 16 July 1964. If a grant were made of a future easement by *A* to *B* the period during which one would 'wait and see' whether the grant were saved by taking effect in possession would be governed by the lives of *A* and *B* plus twenty-one years. There is scanty and conflicting authority on what constitutes a future grant and what constitutes an immediate grant. *Sharpe* v *Durrant* (1911) 55 SJ 423 should be contrasted with *South-Eastern Railway Co Ltd* v *Associated Portland Cement Manufacturers (1900) Ltd* [1910] 1 Ch 12. In every case, as a precaution, the grant which has a future element should be limited to the perpetuity period. There is further doubt about whether a grant of a right to use pipes and sewers 'now laid or hereafter to be laid' under the servient land and not limited to the perpetuity period is completely valid, completely invalid, or valid as to those laid at the date of the grant and invalid as to those subsequently laid or to be laid. Can the parties be assumed to have intended that if their grant was not completely effective it was to be partly effective? Cross J said in *Dunn's* case, above, at 440, that if there had been existing sewers at the date of the grant that would have conferred an immediate right to make connections to them in the future, under the *Associated Portland Cement* case. This was an obiter dictum of Cross J. If there were no existing sewers at the date of the grant then the grant would be wholly in the future and should certainly be limited to the perpetuity period. By s 1(1) of the 1964 Act, where an instrument made after the coming into force of the Act so provides, the perpetuity period may be of such duration not exceeding eighty years as the instrument specifies in that behalf.

The interpretation put on s 162(1) of the Law of Property Act

1925 by *Dunn's* case must be borne in mind. The subsection says that the rule against perpetuities does not apply and shall be deemed never to have applied to (inter alia) 'any grant, exception, or reservation of any easements . . . for the purpose of . . . constructing, laying down, altering, repairing, renewing, cleansing, and maintaining sewers, watercourses, cesspools, gutters, drains, water-pipes, gas-pipes, electric wires or cables or other like works' (para *(d)*(iv)). Cross J said in *Dunn's* case (at 441) that if there was a valid right over property not itself void for perpetuity, ancillary rights of constructing to make the main right effective were not to be void merely because they might be exercised outside the perpetuity period. So the section will save only *ancillary* rights, under the current judicial interpretation. A precedent for connecting into drains on a building site in the future (period limited to 20 years) is given in *Precedents for the Conveyancer*, prec 16–23 p 8099, and for mutual rights on two building plots being created out of a larger one *see* prec 19–37, p 9642.

Eavesdrop. Another type of pipe easement is the right to drain the rainwater from premises onto the land of another, by a downspout or gutter. It is prudent for the servient owner to stipulate the extent of the grant precisely. He should say where and how the drainage is to take place. He should also state what is to be drained: if the intention is to permit the drainage of *A*'s outhouse roof onto *B*'s land, the grant should say so expressly and not mention *A*'s premises in general, otherwise *A* could later validly add the drainage from the whole of his roof provided the drain would accommodate it. An express grant without limiting words will not be limited to the use contemplated when the grant was made (*British Railways Board* v *Glass* [1965] Ch 538).

The grant should make provision for the repair of damage caused to *B*'s servient premises by rainwater from the dominant premises, whether by the natural process of drainage or by overflow. *A* in turn will require a right of access onto *B*'s land to inspect, cleanse and repair the pipe or gutter and will be made responsible for any damage caused in the process of exercising his right. The *Encyclopaedia*, 5th ed, Vol 12, p 1269 gives precedents.

1.4 Licences

Pipes may be laid in or on someone else's land by virtue of a licence. The nature of licences and their assignment is discussed in Chapter 2/2 (a), (b). A precedent may be found in *Precedents for the Conveyancer*, 1, 3034.

The grant of a licence can be similar to the grant of an easement and in drafting the licence it will be useful to bear in mind the considerations relevant to a grant of easement. What is the land to be benefited and the land to be burdened? What are the purpose, the extent and the time of beginning and end? What will the pipe be made of and what materials may be passed down the pipe? Who will repair and eventually remove the pipe and reinstate the land? Is the licence determinable and on what grounds? In the absence of stipulation to the contrary, pipes laid in land become the property of the person in whose land they are laid (*Armstrong* v *Sheppard & Short Ltd* [1959] 2 All ER 651).

It may be useful to incorporate expressly s 196 of the Law of Property Act 1925 (as amended by the Recorded Delivery Service Act 1962, s 1(1)), which deals with the service of notices, although by s 196(5) the section applies to 'notices required to be served by any instrument affecting property' coming into operation after 1 January 1926, unless the contrary appears. Under the section, a notice sent by registered post or recorded delivery service to the person to be served is deemed duly served if not returned through the post office undelivered.

There can be due service even if the addressee does not receive the letter. In *Re 88 Berkeley Road, London, NW9* [1971] 1 All ER 254 one of two joint tenants caused a notice of severance of the joint tenancy to be sent to the other. The notice was received at the premises by the person who had caused it to be sent and the other claimed not to have seen it. There was held to have been due service on the other.

2 Acquisition by implied grant or reservation

Lord Parker of Waddington said in *Pwllbach Colliery Co Ltd* v *Woodman* [1915] AC 634, at 646–7, that apart from ways of necessity and continuous and apparent easements (both dealt with later in this section) there are two heads of easements granted by implication. The first is where the implication arises because the right in question is necessary for the enjoyment of some other right expressly granted. He gave the example that a right to draw water from a spring necessarily includes the right to go on someone's land to the spring for this purpose. He added that the second case depends not on the terms of the grant but on the circumstances under which it is made: the law will imply the grant or reservation of such easements as may be necessary to give effect to the common intention of the parties with reference to the manner or purposes in and for which the land granted or some land retained by the grantor is to be used.

2.1 Implied ancillary rights

The grant of an easement is prima facie also the grant of such ancillary rights as are reasonably necessary to its exercise or enjoyment (*Jones* v *Pritchard* [1908] 1 Ch 630, per Parker J, at 638). The same proposition applies to a right gained not by grant but by prescription. In *Roberts* v *Fellowes* (1906) 94 LT 279, a person who had a prescriptive right to a flow of water to his mill over the defendant's land, along a stream the banks of which had been artifically built up to maintain the flow, was held entitled to enter onto the defendant's land to repair the banks when they were undermined by rats and moles. A grant of a right to lay pipes in another's land carries with it the right to enter on the land to repair the pipes: *Pomfret* v *Ricroft* (1669) 1 Wms Saund 321 (6th ed), at 323 (explicable also, however, on the principle of non-derogation from grant). Indeed, there is a duty on the owner of the pipe easement to keep his pipes in repair: *Humphries* v *Cousins* (1877) 2 CPD 239 (occupier of house held liable when his drain, which ran under a neighbour's public-house, leaked and caused damage there). A right to pass and repass on foot along a way does not carry an ancillary right to lay flexible pipes along the way to empty a cess-pit of sewage (*Penn* v *Wilkins* (1974) 236 EG 203).

Many modern property developments include projecting upper level pedestrian footways. If the developer does not own the land over which the pedestrian way extends, he will need a grant of an easement to project into the air-space which it occupies. It is suggested that he will by implication (if it is not granted expressly) acquire ancillary easements to lead drainage pipes down to ground level and to enter onto the ground floor land when required to repair the elevated way and its supports and pipes.

Has a pipe laid in land a right of support and protection from the soil in which it is laid? In *Newcastle-under-Lyme Corporation* v *Wolstanton Ltd* [1947] Ch 427 (a case where pipes were laid under statutory authority) it was held that the pipe-owners gained no right of ownership in respect of the soil in which the pipes were laid but they had a right to occupy that cavity which their pipes filled; their only right to support was an implied one (which in this case had been taken away by statute). Where the pipes are laid by private grant or under prescription it is suggested that the dominant owner will be entitled to have his pipes supported by the land over which they run (and protected by the soil if the nature of the pipe and its contents require it), or if not by the land then by an adequate substitute. Support and protection are ancillary rights necessary to the enjoyment of the pipe easement (*Jones* v *Pritchard* [1908] 1 Ch 630, at 638). A person has no right apart from agreement to have his pipes supported by the water which is under the land in which the pipes run (*Popplewell* v *Hodkinson* [1861–73] All ER Rep 999), but he can object if the support afforded by wet sand (*Jordeson* v *Sutton, Southcoats and Drypool Gas Co* (1899) 2 Ch 217) or brine (*Lotus* v *British Soda Co Ltd* [1972] Ch 123) is removed.

The rights to repair pipes and to replace with other pipes when deterioration makes it necessary are obviously ancillary to the grant to the dominant owner. The right to substitute an improved type of pipe or cable will be included in the right to lay. All these will require access and, provided the disturbance caused to the servient owner is not unreasonable and is made good, he will not be entitled to complain. Indeed, if the dominant owner does not keep his pipe in repair, so that water escapes onto the servient land, he will be liable in trespass to the servient owner (*Jones* v *Pritchard*, above, at 638; *Humphries* v *Cousins*, above).

By contrast, apart from any special local custom or express contract, the owner of a servient tenement is not under any obligation to the owner of the dominant tenement to execute any repairs necessary to ensure the enjoyment of the easement by the dominant owner (*Duke of Westminster* v *Gould* (1983) 267 EG 762 (CA)). But where the defect on the servient tenement amounts to a nuisance the servient owner is liable to the dominant owner if he does not take reasonable steps to abate it: *Leakey* v *National Trust* [1980] QB 485 (CA) (landslip); *Bradburn* v *Lindsay* [1983] 2 All ER 408 (dry rot and disrepair).

The owner of land over which there is a right of way owes no duty of care, at common law or under the Occupier's Liability Act 1957; persons lawfully using the way are not necessarily the occupier's 'visitors' (*Holden* v *White* [1982] 2 WLR 1030 (CA)).

It is a question of degree in every case whether the right claimed is reasonably ancillary to the right granted. For example, if a person is granted the right to construct a tunnel through his neighbour's embankment to reach the main road from his own land and he uses the tunnel as a garage for his vehicles, he could be restrained. Parking for a brief period whilst he shuts the gate from the main road, for example, would be ancillary to the grant; garaging would not (*Bulstrode* v *Lambert* [1953] 1 WLR 1064, at 1071).

Where there is a right to have a pipe running through another's land there is an ancillary right with it to prevent him from doing anything on his land which will make the exercise of the ancillary right of repair substantially more difficult than formerly (*Goodhart* v *Hyett* (1884) 25 Ch D 182).

2.2 Common intention of parties: implied grant or reservation

Wilde B said in *Hall* v *Lund* (1863) 1 H & C 676, at 685–6, that in cases of implied grant the implication must be confined to a reasonable use of the premises for the purpose for which, according to the obvious intention of the parties, they are demised; each case must depend on its own circumstances and the intention of the parties to be ascertained from the character, state and use of the premises at the time of the grant. For example, in *Ewart* v *Cochrane* (1861) 5 LT 1, a house adjoined a tan yard and both were owned by the same person. The tan yard was drained by a gutter into a cesspool in one corner of the garden of the house. Then the tan yard was sold off and the system of drainage continued. When the house was sold its purchaser sought to prevent the drainage. It was held that the right to use the drain became an easement on severance and passed by implied grant as it was necessary for the convenient and comfortable enjoyment of the tan yard. (This does not mean that there was an 'easement of necessity'. 'Necessity' has a narrower meaning than 'necessary': *Union Lighterage Co* v *London Graving Dock Co* [1902] 2 Ch D 557, per Stirling LJ, at 573).

An implied easement may arise where properties are severed by being devised by will to separate persons. In *Schwann* v *Cotton* [1916] 2 Ch 459, three adjoining properties, *A*, *B* and *C* were served with water by a pipe running underground from a well on *A*. The testator, who owned *B* and *C*, devised them to different persons, who in turn sold to Cotton and Schwann respectively. Cotton stopped up the flow of the pipe to *C* and Schwann claimed successfully an implied right to have the flow continued.

(i) The rule in Wheeldon v *Burrows*. Where the ownership of

adjoining premises is severed the rule in *Wheeldon* v *Burrows* (1879) 12 Ch D 31 is frequently important. It was in this case laid down that on a severance of ownership the grantee will obtain by implication all those continuous and apparent easements which are necessary to his reasonable enjoyment of the property granted *and* which were enjoyed by the grantor (as quasi-easements) before severance. *Ewart* v *Cochrane*, above, though decided before *Wheeldon* v *Burrows*, is an example of the application of the principle. In *Watts* v *Kelson* (1871) 6 Ch App 166, a person who had two pieces of land made a water supply to the second from a tank on the first. He sold off the second, and then the first. It was held that the owner of the second could insist on the continuance of the supply from the first. It will be a question of fact in each case what easements are reasonably necessary, whether they have been 'continuous and apparent' (ie, evidences of their existence can be seen on inspection) and whether they were enjoyed by the vendor at the time of severance. The fact that a pipe was temporarily blocked at the time of severance or had burst in a recent frost would not prevent the right to use it from passing by implication. The case would be otherwise if it had been blocked or burst many years earlier and had not been used since. No right will pass by implication under the rule in *Wheeldon* v *Burrows* if the grantor was incapable of making an express grant of the easement in question. In *Re St Clements, Leigh-on-Sea* [1988] 1 WLR 720 (incumbent of church unable to grant claimed right of way without a faculty from the ecclesiastical court).

For the effect of the Law Society's Conditions of Sale (1984 revision) and of the National Conditions (20th ed 1981) on the grant or reservation of easements on a sale-off of land, see (iii) below.

(ii) Implied reservation. *Wheeldon* v *Burrows*, above, will not imply reservations in favour of a grantor; if the vendor wishes to reserve any rights out of the property to be conveyed he must do so expressly (Thesiger LJ, at 49; *Cordell* v *Second Clanfield Properties Ltd* [1969] 2 Ch 9). So if *A* grants a right to *B* to lay and use a drain under *A*'s land, *A* cannot then grant to *C* a right to tie into and use the drain laid by *B*, unless *A* reserved against *B* the right to make such a grant (*Simmons* v *Midford* [1969] 2 Ch 415). The only case in which an easement may be reserved in favour of a grantor is where he can claim an easement of necessity.

To do this he will have to establish that without the easement the property retained cannot be used at all, not merely that inconvenience would be caused to him if he did not have it (*Union Lighterage Co* v *London Graving Dock Co* [1902] 2 Ch 557, at 573, per Stirling LJ). An obvious example is where the owner sells some land and leaves himself with an insufficient way to carry on his existing business (*Gayford* v *Moffatt* (1868) 4 Ch App 133). But rights to a supply of water and to drain away foul water over the property sold are equally important and it appears that they would be held to be easements of necessity. In *Williams* v *Usherwood* (1983) 45 P & CR 235 (CA) it was held that where an adverse possessor established a possessory title to the surface of land he could be subject to the rights of the paper owner, which had the practical characteristics of easements, as necessities. Thus the adverse possessor gained title to the former dominant land but remained subject to the former owner's right to go onto it for maintaining his property and the right to continue to use the drains under it, which led into the drains common to both properties. In *Wong* v *Beaumont Property Trust Ltd* [1965] 1 QB 173 (CA), a claim to fix a ventilation duct on the servient land, without which duct the public health authorities would close the kitchens of the dominant premises (a restaurant), was held to be a valid claim to an easement of necessity. So if the 'drainage' of smells and fumes could be upheld, so could any other essential drainage claim. It was held in *Holmes* v *Goring* (1824) 2 Bing 76, that a way of necessity does not survive the necessity which gives rise to it and will cease if the person who has a way of necessity to his property acquires another access to it. The decision was doubted in *Proctor* v *Hodgson* (1855) 10 Exch 824.

The Court of Appeal in *Nickerson* v *Barraclough* [1981] 2 WLR 773 disagreed with the view that easements of necessity are based upon public policy. They are based upon implication from the circumstances at the time of the grant. Public policy has no part to play in construing the instrument, though it may require the court to frustrate the parties' intentions when they have been ascertained. If it is clear that without the implication of an easement land is unusable or of little use, that may be an indication that the parties intended it to have an easement, such as one of drainage or ventilation, over the grantor's adjoining land. A declaration excluding or limiting the grant of an easement will not be overruled on the ground that the land could be more fully or more profitably used if the exclusion were removed. An easement of necessity will only be implied into a conveyance of land where without such an easement the property retained could not be used at all. It will not be allowed simply where it is necessary to the reasonable enjoyment of the property, *M R A Engineering Ltd* v *Trimster Co Ltd* (1988) 56 P & CR 1 (CA).

Where the easements claimed are complex and would require to be discussed and negotiated they are unlikely to be allowed by the court under the doctrine of necessity (*Sovmots Investments Ltd* v *Secretary of State* [1979] AC 144 (HL); claim for series of easements for maisonettes in a large tower block development).

Where there are drains used in common by adjoining houses and running under both and the ownership is severed, each house will acquire the right to use the drains running under the other, but will be subject to the similar rights of the other (*Pyer* v *Carter* (1857) 1 H & N 916). This situation only arises where the drainage from *A* runs onto *B*, joins with the drainage from *B* and then comes back under *A*.

When the owner of two properties sells them at the same time to different purchasers who know of the contemporaneous sales, then each acquires the continuous and apparent easements over the other's land which existed as quasi-easements before severance (*Allen* v *Taylor* (1880–81) 16 Ch D 355). Rights to drainage and to a supply of water are examples of such mutual rights.

(iii) Law of Property Act, s 62. Section 62(1) of the Law of Property Act 1925 says that a conveyance of land shall be deemed to include and shall convey (inter alia) all ditches, waters, watercourses, liberties, privileges, easements, rights and advantages whatsoever appertaining or reputed to appertain to the land at the time of the conveyance. By s 62(2) a conveyance of land, having houses or other buildings thereon, shall be deemed to include and shall convey with the land (inter alia) all cisterns, sewers, gutters, drains, watercourses, liberties, privileges, rights and advantages whatsoever appertaining or reputed to appertain to the land or buildings at the time of the conveyance. By subs (4) the section will only apply to the extent that a contrary intention is not expressed in the conveyance, and by subs (6) the section applies only to conveyances made after 31 December 1881. 'Conveyance' is widely defined in s 205(1)(ii). The effect of s 62 is that on a conveyance rights corresponding to the benefits enjoyed before the conveyance can come into existence although they would not be deemed to arise under the rule in *Wheeldon* v

Burrows, above, because that rule requires the enjoyment of the quasi-easement before severance to have been continuous and apparent or necessary to the reasonable enjoyment of the land granted. It is a requirement under s 62 that the right claimed should have been enjoyed with the land conveyed at the time of the conveyance. Section 62 and the rule in *Wheeldon* v *Burrows* can only apply where the conveying party owns the quasi-servient tenement at the time he conveys away the quasi-dominant tenement, *M R A Engineering Ltd* v *Trimster Co Ltd* (1988) 56 P & CR 1 (CA). An example of the wide operation of s 62 is provided by *Goldberg* v *Edwards* [1950] Ch 247, where a person who had agreed to take a lease and had gone into possession was allowed to use a passage to a particular exit as an alternative to the usual passage. When the lease had formally been granted he successfully claimed that under s 62 he had a legal right of way along the passage. The section will not operate to grant rights in property unknown to the law, such as a right to take water from an artificial watercourse whenever the owner of it chooses to let water flow along it (*Burrows* v *Lang* [1902] 2 Ch 502), or a claim to personal services (*Regis Property Co Ltd* v *Redman* [1956] 2 QB 612), nor will it convert into a right a permission intended when granted to be only temporary (*Wright* v *Macadam* [1949] 2 KB 744 (CA); *Green* v *Ashco Horticulturist Ltd* [1966] 1 WLR 889).

In *Sovmots Investments Ltd* v *Secretary of State* [1979] AC 144 (HL) it was held that no easements could be implied under s 62 in respect of maisonettes forming part of a tower development when the maisonettes had not been occupied before the local authority acquired them, as no rights appertained to them before severance. The effect of this decision is limited by the Local Government (Miscellaneous Provisions) Act 1976, s 13 and Sched 1, which provides from 14 February 1977 that on a compulsory acquisition of land a local authority may acquire 'new rights' not in existence when the order specifying them is made. The authority declares what the rights are to be and at the same time acquires them.

In *Nickerson* v *Barraclough* [1981] 2 WLR 773 (CA) it was held that where a way passed under s 62(1) for agricultural and sporting purposes it could not later be used for building purposes when the owner of the easement acquired further land adjacent to the original dominant tenement.

Necessary ancillary rights will be implied. In the case of *Morris* v *Cartwright* (1963) 107 SJ 553, a person to whom a right to a supply of water from adjoining land had passed under s 62 was held entitled to the ancillary right to go onto the servient land to repair the system of pipes to continue his supply.

Where land is sold off under the terms of the Law Society's General Conditions of Sale (1984 ed) and the vendor retains land which adjoins or is near to the land sold, by condition 5(3)*(b)*,

> The conveyance of the property shall contain such reservations in favour of the retained land and the grant of such rights over the retained land as would have been implied had the vendor conveyed both the property and the retained land by simultaneous conveyances to different purchasers.

This imports the rule applicable to simultaneous sales-off under *Swansborough* v *Coventry* (1832) 2 Moo & S 362 and gives the vendor benefits which he would not obtain otherwise at common law, because the view was taken in *Wheeldon* v *Burrows* (1879) 12 Ch D 31 that the vendor acquires no rights by implication on a sale-off and must expressly reserve those he wishes to have.

Where the sale-off is under the National Conditions of Sale (20th ed 1981) the position is more onerous against the purchaser. By condition 20 he does not acquire 'any right to light or air over or in respect of any adjacent or neighbouring property which is retained by the vendor' and the vendor becomes entitled to 'such easements and rights' as would have been acquired under a simultaneous conveyance, as under *Swansborough* v *Coventry* above. Thus s 62 will give the purchaser rights of drainage over the vendor's retained land, but not rights to light or air. An exclusion clause of this type was held to be effective in *Squarey* v *Harris-Smith* (1981) 42 P & CR 118 (CA). The Unfair Contract Terms Act 1977 applies where the contract is made on or after 1 February 1978. The Law Society and National Conditions are commented on in H W Wilkinson, *The Standard Conditions of Sale of Land*, 4th ed (Longman 1989).

(iv) Easements supplied by statute. Where a tenant purchases the freehold under the provisions of the Leasehold Reform Act 1967, the conveyance to him has by implication the effect of granting with the premises (inter alia) rights of passage of water, gas or other piped fuel over adjoining land, and of drainage, so as to secure to the premises the same benefits as were enjoyed under the tenancy (s 10(2)(i)). The premises are subject to similar burdens for the benefit of adjoining premises (s 10(2)(ii). The

conveyance must not exclude the operation of the Law of Property Act 1925, s 62 unless the tenant/purchaser consents (s 10(1)).

Where the tenant of a council house purchases it under the 'right-to-buy' provisions of the Housing Act 1985, Sched 6 sets out the provisions which are to be incorporated in the conveyance of the freehold, or the grant of the lease, as the case may be (s 139). In certain cases the Housing Act 1988 s 79 and Sched 11 (disposal of land by a housing action trust) may also be relevant.

By para 1 of Pt I of Sched 2 of the 1985 Act, the conveyance or grant 'shall not exclude or restrict the general words implied under s 62 of the Law of Property Act 1925 unless the tenant consents or the exclusion or restriction is made for the purpose of preserving or recognising any existing interest of the landlord in tenant's incumbrances or any existing right or interest of any other person'.

By para 2 the conveyance or grant shall have the effect of granting with the house all the rights which the tenant previously enjoyed from adjoining property but make him subject to all such easements and rights for the benefit of other property as are capable of existing in law and are necessary to secure to the persons interested in the other property the same rights as were available before the sale. These rights are rights of support; access of light and air; passage of water, gas or other piped fuel, the drainage or disposal of water, sewage, smoke or fumes, or to the use of or maintenance of pipes or other installations for such passage, drainage or disposal; and rights to the use or maintenance of cables for the supply of electricity, telephone 'or the receipt directly or by landline of visual or other wireless signals' (para 2(1)). Where a lease is granted there are to be covenants by the parties, including one of the landlord to keep in repair structure and exterior (including drains, gutters and external pipes) and another to maintain services at a reasonable level and to keep in repair the installations for the provision of those services (para 13(1)). Where the property sold is a flat, there are covenants implied by the landlord to keep the structure and exterior (including drains, gutters and external pipes) in repair, para 14.

A standard conveyance for a sale under the 1980 Act (now the 1985 Act) is given in *Precedents for the Conveyancer* 2, 8141.

Under the London Building Acts 1930 to 1982 adjoining owners have the right (for example) to underpin, demolish or repair party walls and to enter onto neighbouring land to do so and necessary dealings with pipes and cables would no doubt be ancillary to this right. However the due procedures as to the giving of notice must be given, otherwise the work can be restrained by injunction, *London and Manchester Assurance Co Ltd* v *O & H Construction Ltd* (1988) EGCS 41, applying *Anchor Brewhouse Developments Ltd* v *Berkley House (Docklands Developments) Ltd* [1987] 2 EGLR 173. Bristol has similar local legislation.

In Law Commission Working paper No 151 (1985) on Rights of Access to Neighbouring Land it was suggested that a court order should be introduced which would enable a person to obtain a 'one-off' right of access to neighbouring land to carry out works of preservation to his own land or premises. Works of improvement would not be included except as incidental to preservation. Compensation would be payable to the burdened owner for loss or damage but not for inconvenience or nuisance caused by access. No Report or legislation has followed.

(v) Extent of the implied rights. The extent of the grant will be ascertained from the implication of a reasonable use for the purpose, according to the obvious intention of the parties, for which the premises are conveyed; the character, state and use of the premises at the time of the grant will be a help in ascertaining intention (*Hall* v *Lund* (1863) 1 H & C 676). In this case a lease of a mill was made to Lund. The lessor owned two mills. The one leased was drained through an artificial drain into a natural watercourse which supplied the other mill. The lessee's predecessor, Pullan, had been a bleacher and had drained the waste products from his process into the watercourse. The lease described the premises as 'late in the occupation of Pullan'. Lund was also a bleacher and continued the drainage. The lessor sold the other mill, and the purchaser, who carried on the business of a paper manufacturer, sought to forbid Lund's practice. He was unsuccessful, as it was held that there was an implied grant to Lund to continue to drain his bleaching business as Pullan had done, to the extent that Pullan and he had done before the purchaser bought the lessor's mill. In *Watts* v *Kelson* (1871) 6 Ch App 166, a person was held to be entitled to a supply of pipe-water to his land from other land formerly in the same ownership. The supply had originally been to cattle sheds on the land. The claim was upheld for a supply of water to a house which was built in place of the sheds; granted the existence of the right, the use to which the water was put was irrelevant in this case. But it is suggested that, if a large block of flats had been erected in place

of the cattle sheds and the owner of them had claimed a right to the greatly increased quantity of water which they required, he would have been unsuccessful because the intention to supply such an amount of water could not have been in the minds of the parties at the date of the original grant of the land. The volume required must at some stage so far exceed that originally supplied that a matter of degree turns into a difference in principle. This would certainly be so if the volume demanded would require the installation of a much larger water-pipe. The position would be otherwise if the grant were an express, and not an implied one. An express grant which contains no limiting words will not be restricted to the grant contemplated when the grant is made (*British Railways Board* v *Glass* [1965] Ch 538). The flats in the example would be entitled to as much water as the original pipe could supply; a wide grant might even extend to permit a supply through a substituted larger pipe, as it was held in *Bulstrode* v *Lambert* [1953] 1 WLR 1064 (a case on a right of way) that the wide wording of a grant permitted the dominant owner to remove posts which narrowed the way and to use it thereafter for larger vehicles than had ever used it before. From that adaptation of the way it is not a big step to the enlargement of a pipe. See also *Keefe* v *Amor* [1965] 1 QB 334 (CA).

The easements impliedly granted under an enfranchisement under the Leasehold Reform Act 1967 are such as 'are necessary to secure to the tenant as nearly as may be the same rights as . . . were available to him under . . . the tenancy' (s 10(2)(i)). Rights are also impliedly reserved to the landlord on the same basis (s 10(2)(ii)).

3 Acquisition by prescription

A person who enjoys a benefit which has the necessary characteristics of an easement over the land of another can establish a right by prescription to the easement if he can show a right either at common law under a presumed grant or under the doctrine of lost modern grant or under the Prescription Act 1832.

3.1 Presumed grant

The right at common law depends on the claimant being able to show that he claims under a grant made before 1189. If he can show enjoyment throughout the memory of living persons he will raise a presumption of an ancient grant, as he will also be showing twenty years' user (*Bealey* v *Shaw* (1805) 6 East 208, per Lord Ellenborough CJ, at 215). But his claim can be defeated by proof that the right could not have existed in 1189, for example by showing that a house which claims a right of drainage through the servient land was built on land which was farmland in, say, 1850. The title deeds are obviously vital here. The claim will also be defeated by evidence of enjoyment by licence, perhaps by payment of an acknowledgement rent by the claimant (*Gardner* v *Hodgson's Kingston Brewery Co Ltd* [1903] AC 229); by showing that since 1189 the dominant and servient lands were in the same ownership, thus extinguishing the right by unity of seisin (*Damper* v *Bassett* [1901] 2 Ch 350); by showing enjoyment by force, secrecy or fraud (*Sturges* v *Bridgman* (1879) 11 Ch D 852, at 863); or by showing that the enjoyment was not continuous (*Hollins* v *Verney* (1883–84) 13 QBD 304: the use of a way for hauling timber in 1851–53, again in 1866–68, and again in 1881 held to be insufficient assertion of a continuous right to enjoyment).

3.2 Lost modern grant

'The basis on which the court is entitled to make the presumption of the lost modern grant . . . is that the owner of the allegedly servient tenement or a predecessor in title has, with knowledge of acts which would otherwise be acts of trespass, acquiesced in those acts and, therefore, it must be assumed that the owner of the servient tenement or some predecessor in title of his has given his consent in the proper way, namely by a deed of grant of that easement' (Megaw L J in *Oakley* v *Boston* [1975] QB 271 (CA) at 280). The proof will vary with the right claimed and evidence of user of upwards of twenty years is usually sufficient (*Tehidy Minerals Ltd* v *Norman* [1971] 2 QB 182 (CA)). A claim will fail if it can be shown that there was never a competent grantor (*Neaverson* v *Peterborough Rural District Council* [1902] 1 Ch 557) or grantee. In *Oakley* v *Boston*, above, it was held that an incumbent of a parish had power under the Ecclesiastical Leasing Acts 1842 and 1858 to grant an easement over glebe land but that the consent of the Ecclesiastical Commissioners was needed and the consent could not be presumed in the absence of any evidence of its being granted.

A grant cannot be presumed for an illegal purpose, such as to pollute a river contrary to statute (*Hulley* v *Silversprings Bleaching & Dyeing Co Ltd* [1922] 2 Ch 268). As in the case of a common law claim, force, secrecy, fraud, licence, unity of ownership and interruption can defeat the claim. Involuntary interruption, such as by requisitioning and use under licence from the requisitioning authority, will not affect the claim (*Tehidy Minerals Ltd* v *Norman*, above). Since the user is under lost grant the enjoyment does not have to be 'next before action brought' as in the claim under prescription (see s 4 of the Prescription Act 1832): in *Oakley* v *Boston*, above, a successful claim was made even though user had been continuous from 1927 to 1962 and intermittent from 1962 to the commencement of the action in 1973. An easement may be acquired under the doctrine of lost modern grant even if the dominant and servient owners have the mistaken belief that the enjoyment of the 'easement' is by express grant made when the respective properties were erected, *Bridle* v *Ruby* [1988] 3 WLR 191 (CA) (use of driveway of adjoining house for access to garage, right to use the drive in original conveyance but clause deleted).

3.3 Prescription Act 1832

The Prescription Act 1832 does not alter the common law but it sets out standards by which the effect of the enjoyment of alleged rights can be judged. Section 2 says that no claim at common law by custom, prescription or grant to 'any way or other easement, or to any watercourse, or the use of any water' if enjoyed without interruption for twenty years before the action was brought, shall be defeated by showing only that it was first enjoyed at a time prior to 20 years before the action. The effect of this is that if a person shows 20 years' continuous enjoyment of an alleged right, evidence that no such right was enjoyed 21 years ago will not defeat his claim to an easement. The claim could be defeated by other defences, however, such as secret or violent enjoyment or permission. The section goes on to say that where there is shown 40 years' enjoyment without interruption before the action the right shall be deemed 'absolute and indefeasible, unless it shall appear that the same was enjoyed by some consent or agreement expressly given or made for that purpose by deed or writing'. Again, the user must be as of right to be valid. The section does not apply to claims to a right of light, which must be established under s 3. The peculiar phrase in s 2 'to any way or other easement, or to any watercourse, or the use of any water' seems to imply that a claim to a watercourse or use of water is not an easement, and the phrase 'way or other easement' could be read under the ejusdem generis rule as 'way or other easement in the nature of a way'. It has been said, however, that the section applies to all easements (*Dalton* v *Angus* (1880–1) 6 App Cas 740, per Lord Selbourne LC, at 798). A right to divert a watercourse (*Mason* v *Shrewsbury and Hereford Railway Co* (1871) LR 6 QB 578) and a right to pollute a watercourse (*Wright* v *Williams* (1836) 1 M & W 77) have been recognised under the section.

If the claimant shows user for a less period than the twenty or forty years, no presumption can be made from this alone in support of a claim (s 6) but the court might still find from all the circumstances evidence of an implied grant (*Hanmer* v *Chance* (1865) 4 De GJ & Sm 626). It is an open question whether, if a period of user commences upon an express but verbal agreement inconsistent with prescription, the person enjoying it can after twenty years' enjoyment claim an easement by prescription. In *Reilly* v *Orange* [1955] 2 QB 112 a right of way to premises was granted verbally to last until the grantee should have made an access over his own land to the premises. He claimed an easement by prescription. He failed on insufficient length of enjoyment, so the question put did not arise, but Jenkins LJ, at 118–19, doubted whether an easement could arise contrary to the plain intention of the parties when the permission was given. It is suggested that, as enjoyment under a claim by prescription must be enjoyment 'as of right,' no easement could arise by user under an intended temporary and conditional permission such as was given in this case. Support for this view is provided by *Hughes* v *Griffin* [1969] 1 WLR 23 (CA): there the owner of a house conveyed it as a gift to his nephew in 1951, but remained living there until his death in 1965. His widow claimed that he had re-acquired the title to the property from the nephew by adverse possession: it was held that his possession from 1951 was by licence and so was not 'adverse' and time never began to run. But it is possible for a user which began as permissive to change its character and become adverse, so that time starts to run (*Healey* v *Hawkins* [1968] 1 WLR 1967).

In computing the 20 or 40 year period, any period of interruption which lasts for less than one year is to be disregarded (s 4). The full period of enjoyment must be shown to have passed (*Reilly* v *Orange*, above) and to continue up to the date when the action calling the alleged right into question is commenced (s 4). The commencement of an action concerning the right is not an 'interruption' within s 4 (*Reilly* v *Orange*). In *Newnham* v *Willson*

(1988) 56 P & CR 8 (CA) a person had used a driveway from 1960 until it was deliberately obstructed by the defendants in spring 1983 and on 23 June 1983 the plaintiff's solicitors wrote to protest about the obstruction. The plaintiff commenced proceedings on 27 June 1984 but failed as there had been an interruption lasting more than a year, from March 1983, so there had not been user for 20 years 'next before some suit or action' under the Prescription Act 1832, s 4. If the right claimed has been enjoyed for only 19 years and one day, the objector is too late to interrupt it because his interruption cannot last for one year before the 20 year period is up (*Flight* v *Thomas* (1841) 8 Cl & Fin 231). The claimant of the easement should not make the mistake, however, of starting his action before 20 years' enjoyment has gone by, because he will fail under the plain words of s 2, which require him to show twenty years' enjoyment (*Reilly* v *Orange*).

The prescription period is not necessarily interrupted if the route of the easement being gained is altered during the period of acquisition: *Davis* v *Whitby* [1974] Ch 186 (CA) (one way used for 15 years: another way nearby substituted for it and used for 18 years: valid easement acquired).

Any period during which the person capable of objecting to the user is under disability, being 'an infant, idiot, *non compos mentis, feme covert*, or tenant for life', is excluded from the computation of the prescription period where the claim is for the easement by reason of 20 years' enjoyment, but not where it is a claim by reason of 40 years' enjoyment (s 7).

Where there is 'any land or water upon, over or from which any way or other convenient (a possible misprint for 'easement': *Wright* v *Williams* (1836) 1 M & W 77) watercourse or use of water shall have been or shall be enjoyed or derived', s 8 applies. Under the section, where the alleged servient tenement has been held by a life tenant or for a lease exceeding three years, then the period of the life interest or tenancy is disregarded in computing 40 year enjoyment provided that the person entitled to the reversion commences his action claiming that there is no prescriptive easement within three years of the cessation of the life interest or lease. The section does not apply to a claim based on 20 years' enjoyment, so if there is such a claim the fact that the servient land is leased for part of the time is irrelevant: *Palk* v *Shinner* (1852) 18 QB 568 (a good claim to an easement by 20 years' enjoyment even though the land had been tenanted for all but the last five years of the 20).

No easement can be acquired by prescription against a servient owner who has no power to make express grants, such as a company for which a grant would be ultra vires (*Rochdale Canal Co* v *Radcliffe* (1852) 18 QB 287). Similarly, an alleged dominant owner with no power to acquire an interest in land (*National Guaranteed Manure Co Ltd* v *Donald* (1859) 4 H & N 8), or a vague body of people, cannot prescribe. A tenant can prescribe for the benefit of his landlord as his enjoyment is that of his landlord (*Gayford* v *Moffatt* (1868) 4 Ch App 133; *Pugh* v *Savage* [1970] 2 QB 373 (CA), but not for his own benefit.

3.4 Nature and extent of prescription

The user required to establish an easement must be such as to indicate throughout the statutory period to a reasonable person in occupation of the servient land that a continuous right to enjoyment is being asserted (*Hollins* v *Verney* (1883–4) 13 QBD 304, at 315). So a benefit enjoyed secretly, or by force, or fraudulently over the servient land for the necessary period will not establish a claim to an easement. In *Liverpool Corporation* v *Coghill & Son Ltd* [1918] 1 Ch 307, the claimants discharged borax solution intermittently and usually at night into the Corporation's sewers, unknown to the Corporation, for over 20 years. Their claim to have an easement to do so failed. Enjoyment by permission will be fatal to a claim (*Healey* v *Hawkins* [1968] 1 WLR 1967) unless the permission was limited and had ended. In *Wood* v *Waud* (1849) 3 Ex 748, a person claimed an easement to have water flow through his land along an underground artificial channel. The water was that pumped off to drain a nearby colliery owned by his neighbour. It was held that he had no right to insist on the colliery owner draining his mine through the channel, so his enjoyment was permissive and precarious only. Prescription depends on the presumption or inference of a grant because of the knowledge and acquiescence of the servient owner (*Dalton* v *Angus* (1880–1) 6 App Cas 740, per Fry J, at 773). The knowledge of the servient owner concerning the user may be actual (he observes what is done, he has the means to prevent it and he does not do so) or constructive (he may be presumed to have known of the user, or he has had a reasonable opportunity of knowing, and he was able to prevent it and did not do so): *Lloyds Bank Ltd* v *Dalton* [1942] Ch 466. In *Diment* v *N H Foot Ltd* [1974] 1 WLR

1427 it was held that the fact that a surveyor arranged tenancies and collected rents of a farm was insufficient to fix the owner with knowledge that a way was being used across the farm by an adjoining landowner.

A prescriptive right in excess of his natural rights may be obtained in respect of water flowing in a natural stream by a riparian owner. The natural right is to use the stream water for ordinary purposes connected with the riparian property, such as watering his cattle, and for extraordinary purposes provided these are connected with the landowner's enjoyment of his riparian land and the water is restored to the stream substantially undiminished in volume and unaltered in character (*McCartney* v *Londonderry and Lough Swilly Railway Co Ltd* [1904] AC 301). The use of river water by a riparian owner for spray-irrigation of his crops, a method which took 60,000 gallons a day had resulted in not more than 4 per cent of the water returning into the river eventually, was held to be a non-permissible purpose as against a lower riparian owner in *Rugby Joint Water Board* v *Walters* [1967] Ch 397. Examples of prescriptive rights are a right against a lower riparian owner to pollute the water (*McIntyre Brothers* v *McGavin* [1893] AC 268) and a right to divert water from a lower owner (*Mason* v *Shrewsbury and Hereford Railway Co* (1870–71) LR 6 QB 578).

In an important recent case, *Scott-Whitehead* v *National Coal Board* (1987) 53 P & CR 263 it was held that:

(*a*) a riparian owner has a natural right to have water in any channel which flows through or abuts upon his land in its natural state both as regards purity and quantity;

(*b*) as against a party who is *not* also a riparian owner, this right is not affected by the use made of the water by the riparian owner, even if he uses it for irrigation and not for domestic purposes and even if he does not substantially return the water to the river after use;

(*c*) a riparian owner can acquire a right by prescription to take water of a certain quality from a stream (for example, water of which the chlorine content does not exceed safe levels) and any appreciable worsening of its state will infringe his prescriptive right;

(*d*) a defendant who wishes to rely upon the right by prescription to pollute a stream must show that his pollution at the time complained of does not exceed that committed during the period of his acquisition of the alleged prescriptive right;

(e) a water authority which knowingly admits into the stream quantities of salt which it must know could be harmful to riparian owners' crops owes a duty of care to the riparian owners to warn them and/or to act with reasonable dispatch to divert the water from them.

Even though a person has acquired a prescriptive right to abstract water he still needs a licence, under the Water Resources Act 1963, from the water authority before he may abstract water from a source of supply (s 23(1) and Water Act 1973, s 9). By s 24(2) of the Water Resources Act 1963, an occupier of land which is contiguous to the water at the place where the abstraction is effected may abstract without a licence for domestic or agricultural purposes (but not for spray-irrigation without a licence). A licence is not needed for an abstraction not exceeding 1,000 gallons, if it does not form part of a continuous operation, or a series of operations, whereby in the aggregate more than 1,000 gallons is abstracted (s 24(1)). In *Cargill* v *Gotts* [1981] 1 WLR 441 it was held that s 24(1) did not authorise a person to abstract as much water as he pleased from any one source of supply, provided only that each abstraction did not exceed 1,000 gallons. Large abstractions taken piecemeal might amount to a series of operations and fall within s 23(1). If a person has a licence to abstract under the Water Resources Act 1963 and complies with it, he has a defence to any action brought either by the river authority or a lower riparian owner in respect of his abstraction of water, s 31(1), per Stuart-Smith J in *Scott-Whitehead* v *NCB* (above).

An easement of eavesdrop for water from a house roof to be discharged onto adjoining land can be acquired by prescription (*Thomas* v *Thomas* (1835) 2 Cr M & R 34).

Prescriptive rights can be acquired in respect of artificial watercourses provided that the court is satisfied that the watercourse is a permanent one. If it is constructed for a temporary purpose, such as to be enjoyed during the continuance of a lease, no easement can be prescribed for because no easement could have been intended or expressly granted by the parties (*Chamber Colliery Co* v *Hopwood* (1886) 32 Ch D 549). Length of enjoyment is not conclusive; a flow of water from 1771 to 1836 over a man's land was held not to give him a prescriptive right to have it continued (*Arkwright* v *Gell* (1839) 5 M & W 203). A 'temporary purpose' is one which in the reasonable contemplation of the parties may come to an end; it is a user which is not meant to be a permanent alteration of the face of nature (*Burrows* v *Lang* [1901] 2 Ch 502, per Farwell J, at 508). The purpose served by the artificial watercourse will often indicate whether it is a temporary one. A polythene pipe laid to supply water to a construction site is apparently temporary, to be used only until the building work is done, and so are drains from prefabricated office buildings on the site. A substantial pipe laid carefully and connected to the plumbing system of the building erected on the site is apparently a permanent artificial watercourse.

The extent of the easement prescribed for is determined by the extent of the user over the prescription period. The user which originates the right is also its measure (*Crossley and Sons Ltd* v *Lightowler* (1867) 2 Ch App 478, per Lord Chelmsford LC, at 481). If the claimant cannot clearly formulate the extent of his claim, but can only show a wide variation in the extent of his user over the alleged prescription period, he will fail (*Hulley* v *Silversprings Bleaching and Dyeing Co Ltd* [1922] 2 Ch 268). Where a prescriptive claim has been established but the user has later been varied it is a question of construction in every case whether there has been excessive user which can be prevented. In *Baxendale* v *McMurray* (1867) 2 Ch App 790, a person with a prescriptive right to pollute a stream with foul water from rag washings at his paper mill was held entitled to go on polluting with the washings from a new vegetable fibre as this was in the reasonable and proper course of his process of manufacture, providing the pollution did not substantially increase. But a right to discharge water into a stream will not usually include the right to discharge polluted water; such a right must be expressly proved if it is alleged to exist (*Cawkwell* v *Russell* (1856) 26 LJ Ex 34). No right to pollute to an extent dangerous to public health can be prescribed for because such a right could not lie in express grant or be presumed to be intended (*Blackburne* v *Somers* (1879) 5 LR Ir 1). If a person has a right of drainage for clean water and sends down foul water, the whole right of drainage may be stopped by the servient owner, since he cannot separate the clean water from the polluted water (*Cawkwell* v *Russell*). If a user has been prescribed for to do something at a particular time and place, it may be an excessive user to do the permitted thing at another time and place. In *McIntyre Brothers* v *McGavin* [1893] AC 268 the plaintiffs had a prescriptive right to take water from a stream at a certain place at their works and pollute it in their works, later returning it to the stream. This they could do at weekends. They claimed to take it at a spot a quarter of a mile below their works, at another time of the week, but were unsuccessful.

A riparian owner has no power to object to an alteration in the depth of the water adjacent to his land if it does not threaten to cause damage to his land or interfere with his acknowledged riparian rights or cause a nuisance to the occupier of the land (*Tate & Lyle Industries Ltd* v *Greater London Council* [1983] 2 WLR 649 (HL)).

There has been a right in the nature of an easement recognised by the courts in cases where, although there has been insufficient enjoyment to establish a prescriptive right, the servient owner has allowed the dominant owner to alter his position as if he had an easement and it would be inequitable to allow the servient owner subsequently to deny the right. In *Rochdale Canal Co* v *King* (1853) 16 Beav 630, a mill was built in 1830 close to the canal and pipes were laid, under the supervision of the canal company's officials, from the canal to the mill. The mill-owners used the water for many purposes known to the canal company but after a user of eighteen years the company unsuccessfully tried to forbid the use. The court said that if a man stands by and encourages another to lay out money under the expectation that no objection will subsequently be raised, the law will not allow him then to interfere with the enjoyment permitted. By contrast, in *Bankart* v *Tennant* (1870) LR 10 Eq 141, an understanding that the owners of a copper-works could use water from a canal so long as they were good customers of the canal company was held to give rise to no equitable obligation on the canal company to go on permitting the extraction of water. A person entitled to the benefit of an equitable right of this nature is not in as good a position as a person with a legal easement or with an equitable easement capable of subsisting as a legal interest. He cannot have specific performance to grant him a legal easement. He probably cannot register his right under the Land Charges Act 1972, s 2, as an equitable easement as he has not a positive right but only a valid defence to an assertion that he has no right (*Gale on Easements*, 15th ed, p 80, considers that he can register as an estate contract). The right may be enforceable against a successor in title of the grantor if its existence must have been patent and obvious to him (*Hervey* v *Smith* (1856) 22 Beav 299: right to use two chimneys in adjoining house), provided that the effect of the Land Charges Act 1972 is not to render actual notice irrelevant and to put all notice on the basis of registration. A dictum of Harman J, in *Hollington Brothers Ltd* v *Rhodes* [1951] 2 All ER 578, at 580, suggests that notice can only be by registration; *Emmet on Title*, 18th ed, p 179–80, agrees and discusses the matter fully. Non-registration is not fatal to enforcement where adjoining owners grant rights mutually to each other (*E R Ives Investments Ltd* v *High* [1967] 2 QB 379 (CA)).

The Law Reform Committee was asked by the Lord Chancellor in 1963 to consider the law relating to the acquisition of easements and profits by prescription. The report of the Committee (14th Report, Cmnd 3100), issued in October 1966, discloses a wide divergence of opinion on what form amendment of the law should take. Eight members of the Committee recommended the complete abolition of the right to acquire easements by prescription; the remaining six members recommended that easements should be capable of being acquired by a period of 12 years' prescriptive enjoyment (para 41), but that a system of notional interruption of enjoyment should be introduced (para 64). The Committee suggested unanimously that the Lands Tribunal should be empowered to discharge easements and also to substitute easements in a different place or of a different nature to secure the more efficient use of the servient land where the existence of the easement is preventing it, subject to a right to compensation to the dominant owner (para 97). No legislation has resulted.

The Law Reform Committee's Report was reviewed by the author at (1967) *Modern Law Review*, vol 30, p 189.

4 Ownership of the pipe

Where a pipe is laid in the land of another and enjoyed exclusively by the dominant owner by virtue of an easement, who owns the pipe? If it is laid under a written agreement which expressly states who is the owner, that will settle the question. If there is no such document the general law will apply. In *Simmons* v *Midford* [1969] 2 Ch 415 Buckley LJ, in discussing the legal maxim that whatever is in the soil is part of the soil, said that it concerned things which were previously chattels but had become physically attached to realty. In such a case the chattel may:

(a) lose its character as a chattel and adhere to the realty so as to become part of it for all purposes; or

(b) adhere to the realty whilst remaining a chattel so that the owner may subsequently detach it and re-possess it; or

(c) never lose its character as a chattel.

The facts of the case must be examined to see which of the three solutions is appropriate. In *Goodhart* v *Hyett* (1884) 25 Ch D 182, where apparently the dominant owner had exclusive use of a conduit through the servient land, he was said by North J (at 186) to own the conduit. It is suggested that ownership of the pipe usually remains in the person who laid it: *Pemsel and Wilson* v *Tucker* [1907] 2 Ch 191 indirectly supports this proposition. Where a company had constructed a tunnel in adjoining land and had enjoyed the exclusive use of it, it was held that the company owned the tunnel itself and not a mere easement in the servient land (*Bevan* v *London Portland Cement Co Ltd* (1892) 67 LT 615). The case, therefore, goes beyond the enquiry whether a person who has an easement owns the pipe.

It was held in an old case, *Nicholas* v *Chamberlain* (1606) Cro Jac 121, that where a man had a house supplied by pipes and he sold off the house without the land through which the pipes came, the pipes would pass on the conveyance of the house: the case also says that if he kept the house and sold the land the pipes would go with the house, but the rule in *Wheeldon* v *Burrows* (1879) 12 Ch D 31 would now prevent this unless the pipes were expressly reserved or the house owner could show an easement of necessity (*Gayford* v *Moffatt* (1868) 4 Ch App 133).

The dominant owner has an implied right to enter on the servient land to repair the pipe (*Pomfret* v *Ricroft* (1669) 1 Wms Saund (6th ed) 321, at 323) and can effectively be under a duty to do so, for he has no easement to have a leaking pipe on the servient land. Even if the dominant owner is not obliged to repair, he is liable for trespass or nuisance caused by an escape from the pipe and must therefore keep it watertight to avoid a trespass, *Ingram* v *Morecraft* (1863) 33 Beav 49, Lord Mansfield at 51, 52, *Jones* v *Pritchard* [1908] 1 Ch 638, Parker J. The position appears to be the same even if the servient owner also uses the pipe and could himself repair, *Buckley (R H) & Sons Ltd* v *Buckley (N) and Sons* [1898] 2 QB 608.

Apart from a requirement by statute, contract or custom there is no obligation on the servient owner to make repairs on his land to preserve the enjoyment of the easement (*Bond* v *Nottingham Corporation* [1940] Ch 429) but if a defect on the servient tenement amounts to a nuisance the servient owner is liable to the dominant owner if he does not take reasonable steps to abate it: *Leakey* v *National Trust* [1980] QB 485 (CA) (landslip); *Bradburn* v *Lindsay* [1983] 2 All ER 408 (dry rot and disrepair). In *Duke of Westminster* v *Guild* [1984] 3 All ER 144 (CA) it was held that where a lease made provision for a drain running under the property of both the tenant and the landlords to be repaired there was no need to imply a covenant to repair on the part of the landlords. Neither the express covenant for quiet enjoyment nor the implied covenant for non-derogation from grant could impose on the landlords positive obligations which they would not otherwise have.

Where there is no written agreement and the dominant and servient owners use the pipe jointly it is suggested that the servient owner will own the pipe to the extent that it runs under his land and that he will be entitled to a contribution to the cost of repair from the dominant owner proportionate to the dominant owner's use. Neither must use the pipe in such a manner as to prevent the other from doing so. If the pipe is fractured or blocked solely by the use to which one party puts it he will be solely responsible for repair.

In *Lee* v *Stevenson* (1858) EB & E 512, it was held that where a person granted to an adjoining owner the right to make a sewer in the former's land he could not then join his own sewer to it and use it. The court held that the dominant owner had the exclusive use of the sewer, 'a sewer of his own' (Crompton CJ, at 518). But it is not clear whether he meant that the dominant owner owned the pipe. In *Simmons* v *Midford* [1969] 2 Ch 415 it was held on the facts that a dominant owner there owned and had the exclusive right to use a pipe under the servient owner's land.

Where the pipe is laid in the land of another by licence, and there is no agreement as to ownership, the pipe will be regarded as part of the realty and will belong to the owner of the land through which it runs (*Armstrong* v *Sheppard & Short Ltd* [1959] 2 QB 384 (CA) per Lord Evershed MR at 401).

Even though a person may own the pipe which runs through another's land, he will not own any of the soil surrounding it. His right of occupation of the land only extends to the cavity which his pipe occupies and not to any of the surrounding or supporting soil (*Newcastle-under-Lyme Corporation* v *Wolstanton Ltd* [1947] Ch 427). He may have ancillary to his pipe easement a right of support and protection from the servient land (ibid).

5 Repair obligations

A dominant owner, though he is not bound to keep the subject of his easement in repair (*Taylor* v *Whitehead* (1781) 2 Doug KB 745) will be liable, if, for example, water escapes from his water-pipe running over another's land (*Ingram* v *Morecraft* (1863) 33 Beav 49, 51, 52; Parker J in *Jones* v *Pritchard* [1908] 1 Ch 638).

A servient owner is not bound to make repairs to ensure the continued enjoyment of the easement by the dominant owner. It appears from *Saint* v *Jenner* [1973] Ch 275 (CA) that if the servient owner disturbs the subject-matter of the easement (in this case by putting 'humps' onto a road across the land in order to slow down traffic, causing the road surface to deteriorate) he may be liable in nuisance to the dominant owner whose enjoyment of the easement is interfered with. The case could be regarded as an example of derogation from grant (in reverse). If the dominant owner covenants to pay a contribution to repair of the way and does not do so, *Gale* (15th ed) p 48 suggests that it is an open question whether his right to use determines on his default or is merely suspended until he makes his contribution (citing *Duncan* v *Louch* (1845) 6 QB 904 per curiam). *Gale* also suggests that the stipulation as to contribution is a covenant which could be enforced against the original grantee by the servient owner for the time being.

Chapter 4

Termination of the Easement

1 Termination by statute

An easement may be brought to an end expressly by statute. For example, under the Town and Country Planning Act 1971, s 118 and Pt VI, on the compulsory acquisition of land by the acquiring authority 'all private rights of way and rights of laying down, erecting, continuing or maintaining any apparatus on, under or over the land shall be extinguished, and any such apparatus shall vest in the acquiring authority' (s 118(1)). A right to compensation is given by s 118(4) to the person who suffers loss by the extinguishment. A right to lay and use pipes, cables and drains would be covered by this section. By s 118(2), subs (1) does not apply 'to any right vested in, or apparatus belonging to, statutory undertakers for the purpose of the carrying on of their undertaking'.

An easement may also be extinguished by statute by implication, where its continued existence would conflict with the carrying out of the thing authorised by the statute. In *Yarmouth Corporation* v *Simmons* (1878) 10 Ch D 518 there was statutory authority to make a pier, but a public right of way would have prevented this being done. It was held that the statute by implication extinguished the way. Similar cases might easily arise affecting pipes and drains. But it is suggested that the easement will only be extinguished by implication if its destruction is a necessary inference from the words used by the statute (*Shonleigh Nominees Ltd* v *AG* [1974] 1 WLR 305 (HL)). In any particular case the court may find that the easement continues. In *Bond* v *Nottingham Corporation* [1940] Ch 429 the servient tenement was compulsorily demolished under the (then) Housing Act 1936, but it was held that the local authority must supply to the dominant tenement an equivalent support in substitution for the support formerly supplied by the servient tenement, the dominant owner having an easement. In *Walsh* v *Oates* [1953] 1 QB 578 (CA) an order for stopping up a highway under the Highway Act 1835 was held to extinguish a public right of way over the road but not a private way over it. In *Mann* v *R C Eayrs Ltd* (1973) 231 EG 843, a person had a right of way over some land. The Air Ministry requisitioned the land and forbade him to use the way but they provided another way for him. After the requisitioning ended it was held that he had not lost the right to use again the first way but he had not acquired a prescriptive right to use the second way.

If a person has a private easement over servient land and acquires the same easement by statute, the private easement will merge in the statutory one. So, if the grant by statute is for a limited period only, after that period the private one will not continue (*New Windsor Corporation* v *Taylor* [1899] AC 41).

Where a person's easement is extinguished by statute he should be advised to consider whether the statute expressly or impliedly entitles him to compensation, for example, under the Lands Clauses Consolidation Act 1845 (*Hill* v *Midland Railway Co* (1882) 21 Ch D 143). The courts will not assume that private rights are intended by Parliament to be removed without compensation, for that is simply confiscation (*Wells* v *London, Tilbury and Southend Railway Co* (1877) 5 Ch D 126, per Bramwell LJ, at 130).

2 Termination by express agreement

Since a legal easement is a legal interest in land (Law of Property Act 1925, s 1(2)(a)), and a release of the servient owner from it by a dominant owner operates as a conveyance of a legal interest (ibid, s 205(1)), the release must be by deed (ibid, s 52). A purported release not under seal may still be of evidential value if it can be coupled with acts of cessation of user to show intention to relinquish the easement. If the strict formalities of release have not been complied with but one party has been induced to alter his position or to lay out money on the basis that there is a valid release, so that it would be inequitable for the person inducing the belief to rely on the informality, he will be restrained from doing so (*Poulton* v *Moore* [1915] 1 KB 400). Similarly, a verbal permission to do some act which will destroy or alter the easement cannot be resiled from by the person giving it, after it has been acted on (*Liggins* v *Inge* (1831) 7 Bing 682). For example, if the dominant owner authorises the servient owner to take up a drain and redevelop the servient land he cannot, in the absence of a prior agreement to that effect, claim to have the drain restored later when the redevelopment has been carried out, even though he had an enforceable easement of drainage before. He should, prudently, make a variation by deed with a fresh right of drainage in the new scheme.

The releasing party cannot release an easement to a greater extent than his own interest in the dominant tenement. For example, a tenant for life of the dominant tenement can, under the Settled Land Act 1925, s 58, release the servient land completely from an easement if he has the consent in writing of the trustees of the settlement. If he has not this consent, he can release it for his own lifetime only and it would again be exercisable after his death (see *Davis* v *Morgan* (1825) 4 B & C 8). There appears to be no decision on the question whether a valid release can be made by a competent releasor to a releasee who has only a limited interest in the servient land. If *A*, who owns the fee simple of the dominant land, which is entitled to a water supply through the servient land, makes a deed with *B*, the tenant for life of the servient land, releasing him and the land from the easement, can *A* re-assert his claim when *B*'s interest ends? It is suggested that *A*'s release might be taken as evidence of abandonment and found to be effective on that score.

3 Termination by unity of seisin

If one person becomes both the dominant and servient owner for the same estate or interest in each tenement then the easements formerly enjoyed by him in the servient land will be extinguished. He henceforth enjoys both lands by virtue of his ownership; he cannot have anything in the nature of a right against himself over his own land (*Kilgour* v *Gaddes* [1904] 1 KB 457). If he subsequently sells off one tenement, in the absence of express agreement the rule in *Wheeldon* v *Burrows* (1879) 12 Ch D 31 (see Chapter 3/2 (b)(*i*)), and the Law of Property Act 1925, s 62 (see Chapter 3/2 (b)(*iii*)), will apply. He will not at common law be entitled to an easement over the land sold off unless he expressly reserves it.

If one person acquires both tenements for other than identical estates the easement will be suspended during his tenure of both and will continue upon termination of his joint holding. For example, in *Simper* v *Foley* (1862) 2 J & H 555, the dominant owner in fee simple acquired a lease in the servient land. Later he disposed of the lease and it was held that after disposal he could continue to enjoy an easement over the servient land which he had enjoyed before he took the lease.

Where a person obtains unity of seisin but not unity of possession, for example where he acquires the fee simple of the servient land and the freehold reversion only of the dominant land because it is subject to a lease, it is not decided whether there is extinguishment of easements or not (*Buckby* v *Coles* (1814) 5 Taunt 311, at 315). It is suggested that whilst the lease continues there can be no extinguishment of any rights which the leaseholder has against the servient land, that the reversioner, not being in possession, cannot use any easements and that once the lease terminates the easements must be extinguished by reason of the unity of seisin. Where a person has unity of possession but not unity of seisin (such as where he obtains a lease of both tenements) all easements existing before such unity will be exercisable after it terminates (*Canham* v *Fisk* (1831) 2 C & J 126).

It may be important to distinguish an easement from a natural right of ownership. If a dominant owner in fee simple acquires a servient tenement in fee simple over which he had previously enjoyed a right of drainage through a pipe, then unity of seisin will destroy his easement, and if he subsequently sells off the former servient land without expressly stipulating for his drainage he cannot claim that the easement revives for his benefit. His only course would be to establish an easement of necessity. If, on the other hand, he acquired the servient land through which a natural stream flowed from the dominant land he would not, after severance, need to show any easement to have the flow of the stream continued. If the servient owner extracted the water for purposes unconnected with his riparian ownership the dominant owner could prevent this by virtue of his natural rights as a riparian owner (*Portsmouth Borough Waterworks Co* v *London, Brighton, and South Coast Railway Co* (1909–10) 26 TLR 173).

A person may own both dominant and servient tenements but in different capacities. For example, a clergyman may have the church vested in him as trustee and the adjoining glebe lands vested in him as beneficial owner (*Ecclesiastical Commissioners for England* v *Kino* (1880) 14 Ch D 213). In this case it was held that the easements continued just as if there were separate owners.

4 Termination by implication

An easement will be regarded as terminated if circumstances can be found which indicate an intention on the part of the dominant owner from which a release or abandonment can be implied. In *Tehidy Minerals Ltd* v *Norman* [1971] 2 QB 182 (CA) it was held that abandonment of an easement or profit à prendre can only be treated as taking place when the person entitled to it has demonstrated a fixed intention never at any time thereafter to assert the right himself or attempt to transmit it to anyone else. In *Williams* v *Usherwood* (1983) 45 P & CR 235 (CA) it was held that an easement is abandoned if the owner of it makes it clear that he has a firm intention that neither he nor any successor in title should thereafter make use of it (adopting the words of Buckley L J in *Gotobed* v *Pridmore* (1970) 115 SJ 78). The various circumstances are categorised below.

4.1 Ceasing to use the easement

Mere non-user of an easement will not of itself amount to abandonment. For example, a person with a right to a supply of water from another's land may cease to use the water because he has sold the cattle he used to need it for, but he may intend to buy some more next year and will then resume the use; again a person may not exercise his right of eavesdrop, because no rain has fallen recently, but he will use the right when the rain falls again. To constitute a case of termination by non-user there must be other circumstances besides non-exercise (*Ward* v *Ward* (1852) 7 Ex 838); 'the non-user must be the consequence of something which is adverse to the user' (Alderson B, at 839). The question of whether the easement has been abandoned is a question of fact in each case. Non-user may be explicable because, for example, an artificial stream previously enjoyed changed its course for some years and then returned to its old course, whereupon enjoyment recommenced. In such a case it may be held that there had been no abandonment (see *Hall* v *Swift* (1838) 4 Bing (NC) 381, a case concerning a natural watercourse). Non-user where the dominant owner has been forbidden to use the easement on requisitioning of the servient land is not evidence of abandonment of the easement (*Mann* v *RC Eayrs Ltd* (1973) 231 EG 843; *Tehidy Minerals Ltd* v *Norman* [1971] 2 QB 183 (CA); *Re Yateley Common, Hampshire* [1977] 1 WLR 840). But failure to exercise the right for a long period may place the dominant owner under the obligation of showing how, during his non-exercise, he gave evidence of his continued claim to the right. In a case where a dye-works had a right to pollute a stream but the works had been demolished a person who claimed to recommence the pollution twenty-five years later under the same right was unsuccessful (*Crossley and Sons Ltd* v *Lightowler* (1867) LR 3 Eq 279). It is not so much the period of cesser as the nature of the adverse act and the intention which it indicates which are to be considered (*Hale* v *Oldroyd* (1845) 14 M & W 789).

There is no rule of law that non-user for any particular time must amount to abandonment comparable to the rule that an easement may be acquired by prescription after enjoyment for a fixed period. In *Ward* v *Ward*, above, an easement was held not to have been abandoned though it was not enjoyed during a period of about thirty-eight years, and in *Cook* v *Bath Corporation* (1868) LR 6 Eq 177, a non-user for forty years did not amount to an abandonment. It was said in *R* v *Chorley* (1848) 12 QB 515, that where non-user or an acquiescence in obstruction were relied on as evidence of abandonment it would not be prudent to accept a period of less than twenty years' non-user or acquiescence. So for practical purposes a minimum of twenty years must be shown, but there is no fixed maximum.

If a person loses his land because of the adverse possession of another he will lose also any easements which he enjoys by virtue of his owning that land. In *Williams* v *Usherwood* (1983) 45 P & CR 235 (CA) the owners of two adjoining houses each owned a share of the joint driveway and had an easement of way over the other's part. By various acts of ownership of one owner and inactivity on the part of the other the first gained a title to part of the drive of the second by adverse possession. It followed that the second lost his easement of way over the drive of the first. He did not however lose the entitlement to use the drains under his 'lost' drive as there had been no adverse act against this user.

A user of the easement in part only will not prejudice any future claim by the dominant owner to resume a full use (*Keewatin Power Co Ltd* v *Lake of the Woods Milling Co Ltd* [1930] AC 640, per Viscount Dunedin, at 657).

4.2 Altering dominant tenement or altering burden on servient tenement

Where a person alters the dominant tenement so as in effect to enjoy a different easement from the one he has acquired, or alters his user so that he brings about the same result, it is a question of construction in every case whether he has evinced an intention to abandon the easement (in which case he cannot subsequently resume the enjoyment of it) or whether he has simply made an excessive user (which can be forbidden without affecting his right to enjoy the original easement).

Mere increase in user will not usually amount to excessive user, as is shown by *British Railways Board* v *Glass* [1965] Ch 538 (increased user of land by caravans); *Woodhouse & Co Ltd* v *Kirkland Ltd* [1970] 1 WLR 1185 (increased user of way by dominant owners' customers); and *Cargill* v *Gotts* [1981] 1 WLR 441 (CA) (greatly increased abstraction of water because of new methods of irrigation by crop-spraying).

A person who materially altered the dominant tenement or the user would in effect be prescribing for a new easement. In *Cargill* v *Gotts* (above) Lawton LJ suggested that if part of the farm had been turned into a trout hatchery the use of water for that purpose might be outside the prescriptive easement already acquired to take water for farm purposes generally.

If a person owns a house which is drained through another's land and he demolishes the house, intending to make an allotment of the land, he cannot subsequently build another house and claim his former easement (*Liggins* v *Inge* (1831) 7 Bing 682). The case would be otherwise if he demolished with a view to reconstruction. The easement will be extinguished if there is evidence from the circumstances of intention to relinquish it, though the non-user may not have lasted twenty years (*Moore* v *Rawson* (1824) 3 B & C 332, at 341).

If the dominant tenement is not demolished but merely altered, the intention to relinquish is harder to find. Not every alteration in the dominant tenement will increase the burden on the servient. In *Thomas* v *Thomas* (1835) 2 Cr M & R 34, a person had a right to eavesdrop from pantiles on his building onto neighbouring land. The building caught fire and was repaired with thatch which projected over the neighbour's land four inches more than the pantiles had done: the wall of the building was also raised about three feet. It was held that the alterations in no way affected the easement. A person with a right to a supply of water to his house who has a bath put in for the first time will probably not use so much more water than he did before as to make any appreciable increase in burden. A changed user which does not materially increase the burden cannot be evidence of an intention to abandon the former easement. But if the person entitled to the water supply added a public car-wash to his house he might require so much more water as to amount to an abandonment of his right to have a domestic supply. If the excessive user complained of is severable from the permitted user the easement will not be lost by the dominant owner, but if no severance is possible he may be held to have abandoned his easement. For example, if a person with a right to drain rainwater through a pipe drains foul water through it continuously he can be forbidden to use it at all, for the clean water cannot be separated from the foul water (*Cawkwell* v *Russell* (1856) 26 LJ Ex 34). If a person 'adds on' another dominant tenement to the one which the easement was granted to serve he may be liable for excessive user. In *Bracewell* v *Appleby* [1975] Ch 408 a person with a right of way to plot *A* acquired plot *B* behind it and built a house there. It was held that he had no right to use the way to go to plot *B*, but an injunction to forbid him from doing so was refused and damages were awarded instead. To have granted an injunction would have made the house on plot *B* unusable. (The correctness of this refusal was doubted in *Anchor Brewhouse Developments Ltd* v *Berkeley House (Docklands Developments) Ltd* [1987] 2 EGLR 173.) If a person adopts a different user completely from the one he had an easement for, then he can be restrained from his new user even though it does not impose a greater burden on the servient land than the old one did. Excessive user becomes irrelevant if he has no right to the new use: *Clarke* v *Somersetshire Drainage Commissioners* (1888) 59 LT 670 (right to pour into a stream the washings from a fellmongery: new business of leather-board manufacturers commenced: held, that there was no right to pour into the stream the foul washings from this even though they were no more foul than the others had been).

If a person loses his dominant tenement because of adverse possession of it by the servient owner he will lose also the easement which the dominant tenement enjoys over the servient land. He may, however, retain an easement over the land so lost if there has been no adverse act towards the easement (drainage retained, way lost *Williams* v *Usherwood* (1983) 45 P & CR 235 (CA); similarly *Marshall* v *Taylor* [1895] 1 Ch 641 (CA)).

One dominant owner cannot sue another dominant owner for interference by excessive user with an easement which they both enjoy unless the user substantially obstructs his own enjoyment; if one causes pot-holes in a road used by both this alone will not found an action (*Weston* v *Lawrence Weaver Ltd* [1961] 1 QB 402), and it cannot be construed as an abandonment of his easement by the offender.

In the absence of agreement one dominant owner is not under an obligation to repair the land the subject of the easement. If he fails to repair he may harm himself by making the exercise of his easement more difficult or even impossible, as where he allows a drain to silt up, or he may provide evidence from which abandonment of the easement can be inferred. A dominant owner is liable for his failure to repair if by so failing he comes to exercise a right different from that granted or prescribed for. For example, if he fails to repair his pipe so that it leaks onto the servient owner's land he now exercises a right to run a leaking pipe over the land, which was not the original right (*Jones* v *Pritchard* [1908] 1 Ch 630, Parker J, at 638). The dominant owner can go onto the servient land so as to repair it to allow continued use of the easement but if he goes beyond repair he may afford himself evidence of ownership (*Williams* v *Usherwood* (1983) 45 P & CR 235).

The owner of the servient land is not under a duty to repair so as to enable continued enjoyment of the easement (*Bond* v *Nottingham Corporation* (1940) Ch 429) but if the neglect of the servient land amounts to a nuisance the servient owner will be liable (*Bradburn* v *Lindsay* [1983] 2 All ER 408).

Part II

Pipes Laid under Statutory Powers

Chapter 5

Statutory Bodies and Compulsory Acquisition

1 Statutory bodies

Statutory bodies are created by statute and derive their powers from statute. They are corporate bodies. For example, the Local Government Act 1972 ('1972 Act') sets up county councils and district councils to be the local authorities to replace urban district, rural district, parish and county councils under the Local Government Act 1933: 1972 Act, ss 2, 270. The Water Act 1973 ('1973 Act') sets up regional water authorities to be responsible for (inter alia) the supply of water (s 11), for sewerage and sewage disposal (s 14), for the prevention of pollution of rivers and coastal waters (s 9), and for supervising land drainage (s 19). The Gas Act 1986 (ss 49 and 57) provided that the property and rights of the British Gas Corporation were to vest in a new company, British Gas plc.

An authority cannot act otherwise than in accordance with its powers; it must not act without powers and it must not exceed those powers it has (*AG* v *Fulham Corporation* [1921] 1 Ch 440). The powers may be either directly conferred by an Act or capable of derivation by reasonable implication from its provisions (*Baroness Wenlock* v *River Dee Co* (1885) 10 App Cas 354, at 362–3), or capable of being regarded as reasonably incidental to something for which there is express or implied authority (*AG* v *Great Eastern Railway Co* (1880) 5 App Cas 472, at 478; 1972 Act, s 111). The powers of an authority are either compulsory or adoptive. A water authority must, for example, provide such public sewers inside or outside its area as may be necessary for effectually draining the area: 1973 Act, s 14. By contrast a local authority may make proposals and schemes relating to the provision of accommodation for handicapped persons: 1972 Act, s 195(2).

In the reorganisation of local government of 1972–74 some of the powers of local authorities were re-allocated. The collection of refuse became the responsibility of district councils and London boroughs in England but the disposal of it became the responsibility of county councils and the Greater London Council: in Wales both duties are performed by the district councils: 1972 Act, s 180, Sched 14; Control of Pollution Act 1974 ('1974 Act'), ss 12 and 14. All public sewers and sewage disposal works became vested in the regional water authority by amendments to the Public Health Act 1936, s 20, made by the 1973 Act, Sched 8, para 33. Town and country planning has largely become the responsibility at local level of the county and district councils by the 1972 Act, s 182 and Sched 16; and the Local Government, Planning and Land Act 1980, Sched 34.

2 Compulsory acquisition

Where an authority has power to acquire land by agreement or compulsorily a standard procedure governs the transaction. The procedure is laid down in the Acquisition of Land Act 1981 and the Compulsory Purchase Act 1965. Persons affected must be notified and advertisement is made in a local newspaper. If there are objections a local enquiry may be held.

If agreement is not reached between the parties compensation is assessed by the Lands Tribunal, under the Land Compensation Acts 1961 and 1973. Relevant considerations are not only the type of land taken and its location but also any reduction in the value of the land retained by the owner by reason of its severance from the land compulsorily taken (Compulsory Purchase Act 1965, s 7). Compensation is also payable where land has been injuriously affected by the taking of land nearby, even though none of the land affected has been taken (Lands Clauses Consolidation Act 1845, s 68; Compulsory Purchase Act 1965, s 10). The classes of 'blighted land' in respect of which a claim for compensation can be made have been extended by Pt V of the Land Compensation Act 1973. Not only the actual use but also the prospective use of the land had the order not been made is relevant in assessing compensation for severance (*R* v *Brown* (1867) LR 2 QB 630: agricultural land rendered unsuitable for building in future). The provisions of the Land Compensation Act 1961, Pt II, ss 5–16, must be applied in assessing the compensation. Section 5 of this Act sets out the following rules for assessment:

(a) No allowance shall be made for the fact that the purchase was compulsory.

(b) The land shall be valued on an open market basis as if the sale were by a willing seller.

(c) The special suitability of the land for a particular purpose shall be ignored if that purpose is one to which it could only have been applied under statutory powers or for which there is no market apart from the special needs of the compulsory purchaser.

(d) If the land is used in a manner contrary to law or injurious to health and by that use its value is increased, the increase is disregarded.

(e) Where land is used for a purpose for which there is no general demand (such as a church) compensation is assessed on the basis of the cost of equivalent reinstatement elsewhere. In *Birmingham Corporation* v *West Midland Baptist (Trust) Association Inc* [1970] AC 874 it was held that the compensation is assessed as at the date when the reinstatement might reasonably have commenced and not at the earlier date when the notice to treat was served.

(f) Compensation for disturbance or any other matter not directly related to the value of the land is unaffected by the provisions of rule *(b)* above.

These six rules must be read in conjunction with ss 7 and 10 of the Compulsory Purchase Act 1965 and Pt IV of the Land Compensation Act 1973.

By s 37(1) of the Land Compensation Act 1973 where a person is displaced from land by a compulsory acquisition he is entitled to a disturbance payment, and by s 38(1) the amount is to be equal to *(a)* his reasonable expenses of moving and *(b)* the loss he will suffer because his trade or business on the land is disturbed. It was held in *Prasad* v *Wolverhampton Borough Council* [1983] 2 All ER 140 (CA) that the loss does not have to follow the dispossession in time; it can be loss incurred by the owner in making arrangements when he knew that acquisition was inevitable. The only requirement was that the claimant must establish that the loss was connected to the acquisition. Thus he may claim the expenses if he moves to other accommodation even before he is given notice to treat.

Regulations govern the procedure and the forms used on compulsory purchase (see the Compulsory Purchase of Land Regulations 1982 (SI No 6)), and vary the rates of interest payable after entry on the land (see the current Acquisition of Land (Rate of Interest after Entry) Regulations). The Department of the Environment issues circulars on aspects of compulsory acquisition; for example, on the Award of Costs incurred in Compulsory Purchase Order Proceedings (No 2 of 1987). The circulars may be found in the *Encyclopaedia of Compulsory Purchase and Compensation* (Sweet and Maxwell), Vol 2, part IV.

Chapter 6

Water Supply Before Privatisation

From 1 April 1974, the responsibility for the public supply of water was transferred from local authorities to the water authority for the area (Water Act 1973 ('1973 Act'), s 11). The 1973 Act set up nine regional water authorities in England and the Welsh National Water Development Authority (s 2, Sched 1). Statutory water companies could remain in existence and continue to supply water as agents for the water authorities (s 12). The constitution and procedure of water authorities is now governed by the new s 3 of the 1973 Act (supplied by s 1 of the Water Act 1983).

The objects of the 1973 Act were to end the previous piecemeal treatment of the supply of water, and to provide for the disposal of polluted water and sewage, the use of inland waters for navigation and recreation, and the restoration of the wholesomeness of rivers, by vesting the powers and duties in respect of each matter in the hands of the one authority in each area (s 1). The powers of the water authority have been increased by the Control of Pollution Act 1974 ('1974 Act'). For example, s 31 of the 1974 Act gives control over the pollution of rivers and coastal waters by any poisonous, noxious or polluting matter or any solid waste matter. Department of the Environment Circular No 100 of 1973 explains the changes under the 1973 Act.

1 Duty to supply

The water authority is under a duty to supply water within its area (1973 Act, s 11), and has a power to supply water outside its limits of supply by agreement with other water authorities (s 13). The water authority is not liable for nuisance resulting from a burst water main attributable to the exercise by the authority of its statutory duty to supply water under pressure (*Department of*

Transport v *North West Water Authority* [1983] 3 WLR 707 (HL)). The duty is to supply 'wholesome' water (s 11). In *Read* v *Croydon Corporation* [1938] 4 All ER 631 (a case under a private Act where the duty was to supply 'pure and wholesome' water) it was held that there was a right of action by the daughter of a water rate payer when she contracted typhoid fever from contaminated water supplied by the corporation. In a Privy Council case, *AG* v *Lower Hutt Corporation* [1964] AC 1469, it was held that the adding of fluoride to the public water supply by a local authority in New Zealand was not a breach of an obligation to supply 'pure water'. The duty to supply water is a general one which cannot be enforced by an individual. In *Hey & Croft Ltd* v *Lexden and Winstree Rural District Council* (1972) 223 EG 1099, some developers demanded a supply of water so that they could erect twenty-seven dwellings for which they had planning permission. It was held that they could not insist on a supply; the proper remedy was an application to the Minister, under s 11 of the Public Health Act 1936, as amended by s 28 of the Water Act 1945, so that he could make an order if he thought fit. The local authority is under a duty to ascertain the sufficiency and wholesomeness of the water supplied in its area and to notify the water authority of any insufficiency or unwholesomeness (1973 Act, s 11(2)). The water authority is under a duty to supply water within its area (1973 Act, s 11(1)). It must ensure so far as possible that premises have a mains or otherwise piped water supply (1945 Act, Sched 3, s 29 and Pt X). The authority may not delay unreasonably in making a water supply. A delay of twenty years in supplying 773 holiday chalets with water and sewers was held too long in *R* v *Secretary of State ex p Jaywick Sands Freeholders Association Ltd* (1975) JPL 663. If a supply cannot be taken to within a reasonable distance of premises and if the existing supply is insufficient or unwholesome to the extent of being a danger to health, the water authority must provide a supply other than in pipes (by water tanker, for example) if this can be done at reasonable cost (1973 Act, s 11(3)). If neither of these can be provided at reasonable cost the water authority can require the owners of houses to provide their own sufficient supply of wholesome water for domestic purposes (1936 Act, ss 137, 138). Domestic purposes include drinking, washing, cooking and sanitary purposes (1945 Act, Sched 3, Pt I). Use of a bath having a capacity of more than fifty gallons is not a domestic purpose (ibid), and neither is a supply for a business such as a laundry (ibid). Where there is a piped supply available but a house is not connected to it, the water authority may require connection to be made if this would be reasonable (Building Act 1984, s 69). If it would not be reasonable the authority may require connection to some other source (such as an artesian well), or if this also would not be reasonable it may require the owner to provide a sufficient supply within a reasonable distance of the premises (ibid). If the owner does not comply, the authority may itself provide or have provided a supply of water and may charge the owner with the expenses reasonably involved in so doing (ibid).

The 'premises' which are entitled to demand a supply of water from the authority need not be conventional structures: in *West Mersea Urban District Council* v *Fraser* [1950] 2 KB 119 a houseboat moored on a mud flat was held entitled to a supply. It is questionable whether a person who occupies premises as a trespasser is entitled to demand a supply. In *Woodstock* v *South Western Electricity Board* [1975] 1 WLR 152 it was held that a 'squatter' in premises was not entitled to demand a supply of electricity as he was not an 'owner or occupier' within the Electric Lighting (Clauses) Act 1899, Sched, s 27. The relevant s 69 of the 1984 Act uses the word 'owner'.

By s 30 of the Water Act 1973 (as amended by s 2 of the Water Charges Act 1976) a water authority has power to fix such charges for the services performed, facilities provided or rights made available as it thinks fit. This amendment was made because of *Daymond* v *South West Water Authority* [1976] AC 609 where it was held that a water rate was not payable by a person who is not connected to the water supply; now a reduced rate could be levied. A person who had no water supply to his shop premises but had a contractual right under his lease to use a communal lavatory a short distance away was held liable for a water rate under ss 38 and 59 of the 1945 Act in *West Pennine Water Board* v *Jon Migael (North West) Ltd* (1975) 73 LGR 420 (CA). In *South West Water Authority* v *Rumbles* [1985] 1 All ER 513 (HL) a shop had no water supply or connection to drainage but the roof on the shop premises drained into a sewer. As the shop had the use of the roof and its drainage facility it was liable for a sewerage charge. Section 59 defines 'premises' as including any interest in land and any easement or right in, to or over land.

An express planning permission is not required by the water authority for the laying underground of mains, pipes or other apparatus, as there is a general permission given by the Town and

Country Planning General Development Order 1988 (SI No 1813), Sched 2, Part 17, Class E.

2 Service and communication pipes

An owner or occupier (but probably not a squatter: *Woodstock* v *South Western Electricity Board* [1975] 1 WLR 152) can demand a supply of water for domestic purposes if he provides a sufficient part of a service pipe and pays or tenders a water rate; a service pipe is composed of a section which is laid in the street – the communication pipe – and a section which is on the premises to be supplied – the supply pipe (Water Act 1945, Sched 3, para 1). Where the owner desires to lay the supply pipe he must give the water authority fourteen days' notice of his intention and tender his water rate (Water Act 1945, Sched 3, para 40). A precedent for a notice is given in the *Encyclopaedia*, 4th ed, pp 23, 99. When he has laid his supply pipe the water authority must within fourteen days lay a communication pipe and any part of the supply pipe which is to be in a highway and connect the communication pipe to the supply pipe (Water Act 1945, Sched 3, para 41). The authority is entitled to insist on security for the cost of the work or payment in advance (ibid). The authority is entitled to insist that each house has its own service pipe (Water Act 1945, Sched 3, para 42 and Water Act 1981 s 4), but if two or more houses were supplied by one service pipe on 1 October 1945, then the authority may not require them to be separately served unless the existing pipe becomes defective or inadequate, or payment of the water rate goes into arrears, or the premises are converted into a larger number of houses than they were formerly (ibid). The communication pipes are vested in the water authority, and it pays the cost of maintenance, repair and renewal of them (Water Act 1945, Sched 3, para 44). The authority must maintain, repair and renew any part of a supply pipe which is laid in a highway but at the expense of the owner of the premises served (ibid). A communication pipe is that part of the pipe which is between the water main and any stopcock placed in the premises as near to the street as possible, or in any other case the part of the pipe between the main and the boundary of the street; a supply pipe is that part of a service pipe which is not a communication pipe (Water Act 1945, Sched 3, para 1).

The authority has power to lay in streets such service pipes as are necessary to supply the premises, and to inspect, repair, alter, renew or remove them (Water Act 1945, Sched 3, s 21). There is power to enter onto private land to inspect, repair, alter, renew or remove service pipes lawfully laid there but the authority must pay compensation for all damage caused in so doing (ibid).

3 New premises

Where a local authority is asked to approve the plans for new houses it must not do so unless they show satisfactory arrangements for the provision to them of a supply, sufficient for the domestic purposes of the occupants, of wholesome water from the water authority system, or (if this is not reasonably practicable) from some other piped supply, or (if this is not reasonably practicable) from a source within a reasonable distance of the premises (Building Act 1984, s 25(1)(2) and (6)). If the plans are approved but they are not adhered to and a sufficient supply is not provided, the local authority may forbid occupation until the matter is put right (1984 Act, s 25(3)).

An owner of land who proposes to erect buildings on it for which a supply of water for domestic purposes will be needed can require the water authority to lay the 'necessary mains' and bring water to such a point as will enable the buildings to be connected at reasonable cost (Water Act 1945, s 37). The authority may require certain annual payments from the developer and can also require him to deposit with it a sum not exceeding the total expense of the works, as security (ibid). The question of what are 'necessary mains' under s 37 was considered by the House of Lords in *Cherwell District Council* v *Thames Water Authority* [1975] 1 WLR 448. In that case the local authority wished to have a water supply to some proposed council houses. There was a piped supply nearby which could have served the houses. The local authority was willing to pay the cost of connecting the houses to this supply but the water authority requested in addition a contribution to the cost of a new 27 inch trunk main which was needed to serve the growing demands of the whole area. The local authority refused to pay the contribution. It was held that in the circumstances the trunk main was not a 'necessary main' for the proposed houses and that the local authority need not contribute. In the lower courts the question for consideration had been agreed by counsel to be whether a trunk main is in law capable of being a 'necessary main' [1973] 1 WLR 984 and [1974] 1 WLR 848 (CA)). The House of Lords disapproved of this approach as it was not the question at issue in the case. The Lords held that the word 'necessary' is applicable only to a new main needed for no other purpose than to convey water from a starting point to a point at which it is practicable to connect to premises by a service pipe. In this case there was already a main to which the proposed houses could be connected. By contrast, in *Royco Homes* v *Southern Water Authority* [1979] 1 WLR 1366 (HL) developers had planning permission for the erection of 700 houses on a site and requisitioned the water authority to provide, under s 37(1) of the Water Act 1945, a domestic water supply to them. There was a 4 inch main in a nearby avenue but it was already fully committed and the authority proposed to lay an 8 inch main from a more distant ten-inch main. An 8 inch main was necessary to supply 700 houses. The developers were asked to deposit £115,000 as security for the annual sums falling due under the undertaking but they contended that their contribution should be limited to depositing as much money as would provide a main from the nearest main, the 4 inch one, and that the cost of bringing that main up to a suitable size to supply the site should be on the general body of ratepayers. It was held that: *(a)* the starting point for any main did not have to be the nearest existing main but the point where the new main could be fed with sufficient water, *(b)* what is a 'necessary main' is a question of fact and sound water engineering practice, *(c)* improvements to an existing distribution system at a point before the connection to the new development takes its departure cannot be charged against the developers. The developers were liable for the deposit claimed.

4 Byelaws concerning water

Under the Water Act 1945, s 17, water undertakers may make byelaws to prevent the waste, contamination or misuse of water; the byelaws may regulate the size, nature, materials, strength, workmanship and connecting up of such water fittings as baths, cisterns, closets, taps, valves, meters and pipes (but not mains). Water fittings are defined in s 59. The procedure for the making of these byelaws is in the 1973 Act, s 36 and Sched 7, Pt II.

5 Non-domestic supply

Water undertakers must provide a supply of water on reasonable terms for non-domestic purposes (Water Act 1945, s 27; 1973

Act, s 11(7)). They cannot be compelled to do this, however, if by doing so they would be in danger of not meeting existing or anticipated requirements for domestic purposes without unreasonable expenditure on new capital works such as reservoirs (s 27 above).

It was held in *Staffordshire Area Health Authority* v *South Staffordshire Waterworks Co* [1978] 1 WLR 1387 (CA) that where an agreement for the supply of water at a fixed charge 'at all times hereafter' was made in 1929 the water authority was entitled to terminate the agreement on six months notice in 1975. The parties could not be deemed to have bound themselves on those terms for all time. Similarly (for agreements to dispose of sewage) *Re Spenborough Urban District Council's Agreement* [1968] Ch 139 and *Beverley Corporation* v *Richard Hodgson & Sons Ltd* (1972) 225 EG 799.

6 Street works

Water authorities may break open any street within their area of supply to lay, construct, inspect, repair or renew mains, service pipes, plant or other works (Water Act 1945, Sched 3, para 22; 1973 Act, s 11(7)). 'Street' includes any highway, road, land, footway, square, court, alley or passage, whether a thoroughfare or not (1945 Act, Sched 3, para 1). A main is a pipe laid to give a general supply, as distinct from a supply to a particular consumer (ibid). They may break open streets to lay mains outside their area if they have given notice to the water authority affected and published the proposal in a local newspaper (ibid, para 20). If there are objections a local enquiry will be held, and before the work can be done the proposals must be approved by the Secretary of State (ibid, para 20). If they have power to lay they also have power to break open streets for repairs and renewals. The power to break open a street includes the right of breaking open the roadway of any bridge carrying a street and any sewer, drain or tunnel in or under a street (Sched 3, para 22). This does not allow the structure of the bridge to be interfered with (*Glasgow Corporation* v *Glasgow and South Western Railway Co* [1895] AC 376). Water authorities are subject to the provisions of the Public Utilities Street Works Act 1950 (outlined in Chapter 10/4), and under that Act must give due notice to bodies involved of their intention to carry out work (such as to gas, electricity and fire authorities), take any special precautions required by the nature of the work (such as where they will cause traffic diversions) and make due reinstatement of the road.

Compensation is payable for damage done in the course of carrying out the works (1945 Act, Sched 3, para 22).

7 Works in private land

Where the authority wishes to lay a main in private land it must give reasonable notice to every owner and occupier and obtain the consent of the local authority if the main is to be within 220 feet of any highway (Water Act 1945, Sched 3, para 19, as applied by SI No 293 of 1977, para 7). The consent of the owner, occupier or local authority must not be unreasonably withheld. The Secretary of State will decide whether a refusal of consent is reasonable (ibid; 1973 Act, s 1(2)).

7.1 Compensation for loss and damage

Compensation is payable for damage done to the land or for injurious affection of it (1945 Act, Sched 3, s 19(3)). Examples are manholes or inspection chambers in the land permanently interfering with cultivation, and disturbance to existing crops whilst the main is being laid; it is relevant to consider whether, because of the existence of the main, a possible alternative use of the land such as for house building is totally or partially prevented.

By s 278 of the Public Health Act 1936, as amended by s 14 of the Water Act 1973, a local authority (including a water authority) shall make full compensation to anyone who has sustained damage by reason of the exercise of its powers. In *Leonidis* v *Thames Water Authority* (1979) 77 LGR 722 a water authority, whilst repairing a sewer, blocked the normal entrance to the plaintiff's garage for eleven months and he suffered loss of profit. It was held that he could recover the loss of £2,500. The fact that the authority did not take an unreasonably long time on the job did not relieve them from liability; the plaintiff must not suffer virtual confiscation of his rights. The damage was not too remote. The land had suffered loss during the period because during that time its letting value must have been diminished.

If the works of the authority cause loss of profit to the owner or occupier of land but are wholly carried out on the land of someone else, it does not appear that the owner or occupier can recover his lost profit from the authority, if the acquisition is made under s 68 of the Lands Clauses Consolidation Act 1845, or any

Act which incorporates its terms (*Argyle Motors Ltd* v *Birkenhead Corporation* [1975] AC 99 (HL)). But if the claim is made under some other Act, such as the Gas Act 1986, s 9(3) and Sched 4, it is possible that compensation may be awarded for loss of profit, if *Leonidis's* case is followed. See the discussion in (1974) *Journal of Planning Law*, 257, at 263 (Jones and Wilkinson).

The fact that the authority can enter in future to inspect and repair will affect the value of the land, and the compensation payment will take account of this (*Markland* v *Cannock Rural District Council* (1973) 227 EG 1173). In *Griffiths* v *Swansea Corporation; Morris's Trustees* v *Swansea Corporation* (1960) 11 P & CR 470, the local authority laid a trunk main 3,657 yards long, with one manhole and three valve chambers, in the land of the claimants. The first claimant was awarded £145 in respect of the pipe 1,123 yards long and £17 for the valve chambers and the second £175 for the pipe 2,534 yards long and £16 for the chambers. Both were awarded costs. Any immediate enhancement of value to the land must be set off against the claim. For example, the laying of the main may make the land more readily saleable for housing development if the owner or occupier would have a right to connect to the main.

8 Compulsory acquisition

A water undertaking has power to acquire land by agreement or compulsorily with the consent of the Secretary of State for any of the purposes of its undertaking (Water Act 1945, s 24). Land includes any interest, right or easement in, to or over land (ibid, s 59). The due procedure of notification of persons affected and advertisement in a local newspaper must be followed. If there are objections a local enquiry will be held. The procedure is laid down in the Acquisition of Land Act 1981 and the Compulsory Purchase Act 1965.

Chapter 7

Water Supply After Privatisation

The Water Act 1989 transfers the property, rights, and liabilities of the existing water authorities to a newly-established National Rivers Authority and then to private companies nominated by the Secretary of State. The water authorities are dissolved (ss 1 and 3). Each water authority is required to make a scheme providing for the transfer of its undertaking to the National Rivers Authority after approval by the Secretary of State (Sched 2).

The National Rivers Authority is established by s 1, with a duty to set up regional rivers advisory committees for each area where the Authority considers it appropriate (s 2). The Authority's functions include those of the former water authorities where not assumed by the new water or sewerage undertakings (s 3(1)). A Director General of Water Services is established, with a duty to set up customer service committees, each allocated to a water or sewerage undertaker. The Secretary of State or the Director must exercise his or their powers in a manner considered to be best calculated to secure that the functions of water and sewerage undertakers are properly carried out and properly financed by securing reasonable returns on capital (s 6). In formulating or considering any proposals the Secretary, the Minister, the Authority, the Director and any relevant body must take into account and have regard to conservation of the environment, protection of archeological sites and buildings of architectural or historic interest and the protection of recreational areas of beauty (ss 7, 8 and 9). The Secretary of State may approve codes of practice with respect to environmental and recreational duties (s 9).

The National Rivers Authority and the Director-General of Water Services have regulatory and policing functions.

The Authority is responsible for conserving, redistributing or otherwise allocating water resources in England and Wales (s 120), for entering into water resources management schemes with water undertakers (s 121), for determining minimum acceptable river flows (s 122), for asking the Secretary of State (or the water undertaker may apply) to make drought orders (ss 126–130), for making arrangements about flood defence and establishing regional flood defence committees in place of regional land drainage committees (ss 131–135), for granting licences to abstract water (ss 123, 124), and for monitoring the extent of pollution in controlled waters (s 102(2)).

The Director-General's duties include requiring an undertaker to provide information for enforcement purposes where it appears to him that a company holding an appointment under the Act may have contravened any condition of its appointment or any statutory requirement or other requirement enforceable under s 19; he may publish information appearing to him to be expedient to give to any customer or potential customer of a company holding an appointment (s 34); and he shall make a report each year to the Secretary of State on the matters in s 34(2); he may require one of his customer services committees (set up under s 5) to prepare a report on any matter which he may request (s 35).

1 General duties of water undertakers

By s 36(1) it shall be the duty of every water authority to develop and maintain an efficient and economical system of water supply within its area and to ensure that arrangements are made for providing supplies of water to premises and to persons who demand them; and for maintaining, improving and extending the water undertaking's mains and other pipes. The duty is enforceable by the Secretary of State or the Director (s 36(2)). The Secretary may by regulations prescribe such standards of performance in connection with the provision of supplies as in his opinion ought to be achieved in individual cases (s 37(2)). The Director may order an undertaker to give a bulk supply to another undertaking (s 38(1)).

2 Supplies through water mains

An owner or occupier of premises, a local authority in the undertaker's area, a new town authority, the Development Board for Rural Wales and an urban development corporation are all entitled to require the provision of a water main for any locality and the water undertaker is under a duty to supply it (s 39(1) and (2)) within three months of the date of a requisition (s 39(3)). The undertaker may impose financial conditions on the supplying of a main (s 40).

Sections 41 to 43 deal with the undertaker's duty to make a connection of a service pipe to the water mains. The owner or occupier of land can serve notice on the undertaker requiring a supply of water for domestic purposes and the undertaker will be under a duty to connect at the expense of the person serving notice unless the main is neither a trunk main nor one which is used or to be used solely for the purpose of supplying water for non-domestic purposes (s 42(1) and (2)). Any duty owed under s 42 is owed to the person who served the notice, s 43(1). The duty can include that of laying a service pipe in the street between the main and the applicant's property at the applicant's expense (s 41(3) and (4)) and can be complied with by laying a main and not a service pipe (s 41(5)). The undertaker may impose conditions on agreeing to comply with a notice, such as that financial security is provided for the expenses, that an approved meter is installed or that so much of a service pipe as is to be used by but has not been installed by the undertaker complies with approved specifications (s 42(1)). The undertaker may require the provision of a separate service pipe to any premises which consist of a house or other building or part of a building which is separately occupied and already supplied but does not have a separate service pipe and on receipt of notice from the undertaker the occupier must within three months lay so much of the required pipe as the undertaker is not under a duty to lay (s 49(3)). The undertaker owes a 'domestic supply duty' in relation to any premises connected under s 41 or supplied before the transfer date from the water authority (s 44(1) and (2)). The duty is to provide the premises with a supply of water sufficient for domestic purposes and to maintain a connection between the main and the service pipe, but this is not to apply to a period when it would be reasonable for the water to be cut off or reduced for the purpose of carrying out any necessary works (ss 44(3) and (5), 48).

The owner or occupier of premises can request a supply of water for non-domestic purposes but there will be no duty to supply him if to do so would require the undertaker to incur unreasonable expenditure (s 45). Every water authority at the request of the fire authority shall fix fire-hydrants on its water mains (other than trunk mains) at such places as may be most convenient for affording a supply of water for extinguishing fires in the area and the undertaker must keep the fire-hydrants in good working order (s 46). An undertaker shall supply at the request of a sewerage undertaker a supply of water for such purposes as cleansing sewers and drains, cleansing or watering highways or supplying any public pumps, baths or wash-houses (s 47(1)).

3 Constancy and pressure

By s 50(1) it shall be the duty of a water undertaker to cause the water in such of its mains or pipes as are used for providing supplies of water for domestic purposes or have fire-hydrants fixed on them to be laid on constantly and at such pressure as will cause the water to reach to the top of the top-most building in the undertaker's area. This does not include a duty to supply at a height greater than that to which the water will flow by gravitation through water mains from the service tank or reservoir from which the supply is taken or during any period when it is reasonable for the supply to be cut off or reduced for the purpose of carrying out any necessary works (s 50(2) and (3)). The water undertaker can be liable on summary conviction to a fine for failure to carry out obligations under s 50 (s 50(8)).

4 Quality and sufficiency of water supplies

By s 51(1) it shall be the duty of the water undertaker when supplying water to any premises for domestic purposes to supply only water which is wholesome at the time of supply; and so far as reasonably practicable to ensure in relation to each source of supply, that there is in general no deterioration in the quality of water supplied from that source. Water which becomes unwholesome only after leaving the undertaker's pipes is not to be regarded as unwholesome under s 51(1), (s 51(2)). Regulations of the Secretary of State may make provision for prescribing steps to be taken to ensure compliance with s 51 and in particular may provide for monitoring and recording whether the supply for domestic purposes is wholesome at the time of supply (s 52(1)). The use of non-approved processes, substances and products may be forbidden by regulation (s 52(2)). It is an offence to supply water by means of pipes to premises if the water is unfit for human consumption (s 53(1)). But it is a defence that the undertaker had no reasonable grounds for suspecting that the water would be used for human consumption or took all reasonable steps and exercised all due diligence to ensure that it was not unfit or that it was not so used (s 53(3)). It is the duty of the local authority to take all such steps as they consider appropriate for keeping themselves informed about the wholesomeness and sufficiency of water supplies supplied to premises in their area, including every private supply and to inform the water authority of unwholesomeness or possible unwholesomeness (s 55(1) and (2)). The local authority may serve notice of the unwholesomeness of a private supply on the owners and occupiers of the premises receiving it and on the owners and occupiers of the source of it (s 56); the notice shall not take effect until the period (not being less than 28 days) specified in the notice as the date when it will take effect, and the recipient may make representations to the Secretary of State (s 57). By s 54, where it is not practicable at reasonable cost for a water undertaker to provide a piped water supply for domestic purposes but it is practicable at reasonable cost to supply water otherwise than in pipes and the insufficiency of the supply is such as to cause a danger to life or health and the local authority for the area notify the water undertaker of the danger, then the undertaker must provide a supply other than in pipes; but the local authority will be liable for the charges though they can recover them in whole or in part from the owner or occupier of the premises to which the supply is made. The Secretary of State may by regulation prescribe standards of wholesomeness (s 64), and will in due course meet European prescribed standards.

5 Contamination, waste and misuse of water

By s 60 various offences are constituted in relation to the misuse of water. If an owner or occupier of any premises to which a supply of water is made by the undertaker intentionally or negligently causes or suffers any water fitting for which he is responsible to be or remain so out of order or in need of repair that the water main is or is likely to be contaminated by the return to the pipe of any substance from the premises or the water supplied is likely to be contaminated before it is used or water is likely to be wasted or unduly consumed, he commits an offence (s 60(1)). If a person who is the owner or occupier of any premises supplies any of the water supplied to the premises to another person for use in other premises, or intentionally allows a person to take water for use in other premises unless for extinguishing a fire or to supply an unintentional lack of water there, he shall commit an offence, as shall the user of the improperly supplied water (s 60(2) and (3)). Such an amount as is reasonable for the cost of the misused water can be recovered by the undertaker from the person liable (s 60(4)). Regulations may be made by the Secretary of State for preventing contamination, waste and misuse such as by forbidding the use of non-approved fittings or those made of non-approved materials (s 61(1)). A person designated in writing by a water undertaker or local authority may be empowered to enter any premises for the purpose of ascertaining whether any provision made with respect to any water fittings or the waste or misuse of water is being contravened (s 61(4)) and in order to prevent damage to persons or property may disconnect the service pipe in an emergency or serve notice on the consumer allowing him not less than seven days to carry out the work stated in the notice. In the case of emergency or the premises being unoccupied the undertaker may disconnect (s 62).

6 Power to deal with foul water and pollution

Without prejudice to the powers in s 110 and schedule 18, the Authority and every water undertaker shall have power, on any land which belongs to the Authority or the undertaker, to construct and maintain drains, sewers, watercourses catchpits and other works to intercept, treat or dispose of any foul water arising or flowing upon that land or of otherwise preventing the pollution of any waters whether surface or underground from which the Authority or the undertaker is entitled to take water (s 148). By s 149 the Minister may confer on the Authority or any undertaker such compulsory powers as he considers necessary for enabling engineering or building operations to be done or for discharging water into any inland water or underground strata (s 149). New rights may be authorised to be created and acquired (s 149(4)).

7 Duties to make recreational facilities available

If an Authority or water undertaker is given permission under s 149 to construct a reservoir which may permanently affect the area and is not primarily intended to benefit the inhabitants of that area, the Secretary of State may include in the order provision for facilities for recreation or other leisure-time occupation for the benefit of those inhabitants (s 151(1)). A Code of Practice for recreation and sport on water undertakers' land has been promulgated under s 9.

8 Records of underground works

It shall be the duty of the Authority and the water undertaker to keep records of every resource main, water main or discharge pipe vested in them and of any underground works vested in them and make them available free of charge to the public for inspection at reasonable times (s 158(1)).

It is an offence to recklessly or intentionally interfere with any resource main, water main or other pipe vested in the Authority or any water undertaker or to do so by act or omission or so to interfere with any structure, installation or apparatus as to damage it or affect its use or operation (s 160).

9 Compulsory acquisition

The National Rivers Authority and a water or sewerage undertaker may be authorised by the Secretary of State (or a Minister in the case of the Authority) to purchase compulsorily any land anywhere in England or Wales which is required by the Authority or the undertaker for its purposes in carrying out its functions (s 146(1)). The power includes creating and then acquiring new rights and also extinguishing rights (s 146(2)). By s 147 pipes and sewers laid by the Authority or an undertaker under the provisions of Sched 18 shall vest in the body which laid them. By s 154 nothing in the Act which empowers a body to carry on works shall entitle them to affect injuriously (unless with consent) works or property of certain undertakings listed in s 154(3), being, for example, undertakings of the Civil Aviation Authority, the British Coal Corporation, an undertaking running a telecommunications system, a public gas supplier, the CEGB or Area boards (or their successors), a public utility undertaking, British Railways Board or London Regional Transport. The procedure for compulsory acquisition is contained in Sched 19. The Director may be called upon to investigate a complaint made by an owner of private land in respect of the use by the Authority or a water or sewerage undertaker of their powers; he may direct a sum not exceeding

£1,000 to be paid in respect of the owner's loss, damage or inconvenience (s 155).

10 Water resources

By s 120 it shall be the duty of the National Rivers Authority to take such action as it may from time to time consider necessary or expedient for conserving, redistributing or otherwise augmenting water resources in England and Wales and of securing the proper use of water resources. It shall make water resources management schemes with water undertakers and prescribe minimum acceptable river flows (ss 122 and 125). It shall grant water abstraction licences (ss 123 and 124). The Secretary of State may make general drought orders authorising measures to deal with a shortage of water in an area (ss 126 to 130).

11 Planning consultation

By article 18 of the Town and Country Planning General Development Order 1988, before granting planning permission for development on certain matters the local planning authority is required to consult the water undertaker concerned. Those matters include: development involving mining operations; that involving carrying out works in the bed or on the bank of a river or stream; using land for storing mineral oils and their derivatives; using land for the deposit of refuse or waste; the retention, treatment or disposal of sewage, trade-waste, slurry or sludge (except in the laying of sewers, making pumphouses in a line of sewers, tanks and cess-pits serving single houses or caravans); and the use of land as a cemetery.

12 Repeals

The Water Act 1989 does not (unlike the Electricity Act 1989) repeal substantially the whole body of legislation on which the previous position rested. The main statutes, such as the Public Health Act 1936, the Public Health (Drainage of Trade Premises) Act 1937, the Water Resources Act 1963, the Control of Pollution Act 1974, and the Land Drainage Act 1976 remain, with appropriate amendments where, for example, 'water authority area' is referred to. A number of sections of the Water Acts 1945 and 1973 are repealed and replaced. The following Acts are repealed entirely: The Drought Act 1976, The Water Resources Act 1971, The Border Rivers (Prevention of Pollution) Act 1951.

13 Compensation

In relation to compensation for pipe laying, matters to be considered include payments for a water main easement the payment for which is in practice assessed on so much a metre of run, based on 50 per cent of the vacant possession value of a notional easement strip. In a helpful booklet, '*Sewers and Water Mains*', published by the Country Landowners Association (A5/1986) it is suggested that the minimum width needed to protect the pipe is 10 feet on either side, plus the diameter of the pipe. In addition to payment for an easement there should be a claim for compensation for manholes, valves and other structures whether or not above ground. Under a national agreement made in 1986 British Telecom pay an annual wayleave of £11.47 for an above ground manhole, for loss to agricultural operations. The laying of the pipe will cause damage, disturbance and loss of profits and the professional fees for negotiating all the matters may also be claimed. The British Standards publication, BS 8010: *Pipelines on Land: General*, should also be consulted.

Chapter 8

Sewers and Drains Before Privatisation

1 Construction of sewers and drains

Section 343 of the Public Health Act 1936 ('1936 Act'), defines a drain as 'a drain used for the drainage of one building or of buildings or yards appurtenant to buildings within the same curtilage'. It defines sewers as all sewers and drains used for the drainage of buildings or yards of buildings (except a drain as defined above). So a sewer drains more than one building; many drains often connect into one sewer. A sewer does not necessarily carry sewage. It could carry flood or rain water, for example (*Hutton* v *Esher Urban District Council* [1974] Ch 167 (CA)). A sewer not draining buildings (such as a cesspit at a caravan site) is not within the definition, apparently.

In *British Railways Board* v *Tonbridge and Malling District Council* (1981) 79 LGR 565 (CA) three natural watercourses had been culverted under a railway line. Over the years the surface water from a considerable number of streets and houses had been piped into the watercourses. It was held that the culvert was not a sewer within s 343. The fact that surface drainage from a built-up area had been channelled into the watercourses was insufficient to make the culvert a drain 'used for the drainage of buildings'.

It is a question of fact whether a structure is one building or more than one. In *Hedley* v *Webb* [1901] 2 Ch 126 a pair of semis was held to be one building; in *Humphery* v *Young* [1903] 1 KB 44 each semi-detached house was held to be a separate building. In *Weaver* v *Family Housing Association (York) Ltd* (1975) 74 LGR 255 (HL) pipes from each of a group of terraced houses ran into a common six-inch pipe under one of them. The common pipe was held to be a sewer. In *Cook* v *Minion* (1978) 37 P & CR 58 a pipe connected four or five houses, then it ran into a four-inch pipe which joined the local authority sewer. It was held that the four-inch pipe was not a sewer.

A drain and a private sewer will be the responsibility of a private individual; a public sewer is the responsibility of the water authority. A public sewer is defined in s 20 as reworded by the Water Act 1973, Sched 8, para 33, and includes all sewers constructed by the water authority at its expense or acquired by it and all sewers which a water authority has adopted by a declaration of vesting, under s 17, 1936 Act. The water authority acquires an absolute property in the sewer and the space which it occupies; its interest is an estate in fee simple which will determine if the sewer ceases to be a sewer, and this is a fee simple absolute by virtue of the Law of Property Act 1925, s 7(1) (*Tithe Redemption Commission* v *Runcorn Urban District Council* [1954] Ch 383 (CA)). The rights of any person entitled to use the sewer before the vesting declaration are preserved (1936 Act, s 17(5)); this includes a prescriptive right (*AG* v *Dorking Union Guardians* (1882) 20 Ch D 595). If the water authority does not own the land through which the sewer runs, it acquires a right of subjacent support by implication and the landowner is entitled to compensation (*Re Dudley Corporation* (1881–2) 8 QBD 86). The authority probably also has a right of support from adjacent land, at least so far as the adjacent land is in the same ownership as that occupied by the sewer (*Jary* v *Barnsley Corporation* [1907] 2 Ch 600, at 613).

The authority is under a duty to provide such public sewers, inside or outside its area, as may be 'necessary' for effectually draining the area and to make provision, inside or outside the area, for effectually dealing with the contents of the sewers (1973 Act, s 14(1)).

Although the authority must decide when sewers are necessary it may not defer the decision unreasonably (*R* v *Secretary of State, ex p Jaywick Sands Freeholders Association Ltd* [1975] JPL 663, sewers and water supply not provided for 773 holiday chalets after twenty years, delay unreasonable, mandamus issued).

A breach of s 14 of the 1973 Act by the authority does not give rise to an action by an individual; the duty is not owed to a specific person or group of persons. In *Weaver* v *Yorkshire Water Authority* [1981] CLY 2874 a public sewer passed beneath the plaintiff's house and on occasion became blocked at an interceptor trap, because foreign bodies were put into the sewer by third parties. This flooded the plaintiff's cellar. The sewer was adequate for the plaintiff's house and the other houses which it served. The water authority offered to remove the trap and seal off this exit but the plaintiff refused and claimed an injunction for the authority to divert the sewer. Held (*a*) on the facts there was no breach of s 14, (*b*) no duty was owed under s 14 to the plaintiff or to any person or group of persons. The water authority must endeavour to arrange for the local authority to exercise the sewerage functions of the water authority on an agency basis, Water Act 1983, s 6, replacing the 1973 Act s 15.

A new Sched 4A (in force from 9 May 1983), inserted into the 1973 Act by s 6(2) of the Water Act 1983, concerns arrangements made under s 15 of the 1973 Act for carrying out sewerage functions. By para 2 arrangements made by a water authority are to be made with a view to ensuring that its sewerage functions taken as a whole are carried out efficiently. By para 3(1) all arrangements shall comply with the requirements of such regulations as shall be made by the Secretary of State. By para 4 the duty to make arrangements imposed by s 15(1) shall not apply in relation to sewerage functions in certain cases, such as where the water authority and the relevant authority agree that it would be inexpedient for arrangements to be made in respect of those functions. If the water authority refuses to enter into an agreement where it is of opinion that it would be inexpedient to do so the relevant authority may apply to the Secretary of State for a review of the decision (para 4 (2)). The Water (Sewerage Arrangements) Regulations 1984 (SI No 1788) have been made under Sched 4.

If the authority fails to provide a sufficient sewer it is not permitted to solve its problem by calling upon householders to stop using the sewer, even if the householders originally made a connection into the sewer without permission from the authority (*Fordom* v *Parsons* [1894] 2 QB 780, a case under the similar s 15 of the Public Health Act 1875). But an individual who is caused damage by the authority's failure to provide an adequate sewer can make no claim against the authority (*Clark* v *Epsom Rural District Council* [1929] 1 Ch 287). In *Robinson* v *Workington Corporation* [1897] 1 QB 619 an overloaded sewer flooded Robinson's houses. He failed in his claim for damages. It was held that his remedy was to complain to the Local Government Board under the Public Health Act 1875: the remedy now is a complaint to the Secretary of State, who may make an order against the authority under s 322 of the 1936 Act, as applied to water authorities by s 14(2) of the 1973 Act, compelling the authority to discharge their functions in such manner and at such time as the order directs. Lord Esher said in *Robinson's* case, at 621, that if a duty is imposed by statute and the duty would not exist apart from the statute, and a remedy for default or breach is provided by the statute which creates the duty, there is no other remedy available to an aggrieved person. On the other hand, if the authority tries to remedy a defect or inadequacy of the sewer and does it so negligently as to create a nuisance, an action by the individual who suffers damage will lie (1936 Act, s 31; *Smeaton* v *Ilford Corporation* [1954] Ch 450). The contradictory authorities are analysed in the *Encyclopaedia of Environmental Health* (Sweet & Maxwell) vol 1, paras 1–093 to 1–097. A nuisance brought about by the gradual operation of the elements is not actionable. In *Radstock Co-operative Society Ltd* v *Norton Radstock Urban District Council* [1968] Ch 605, the urban district council laid a sewer under a river bed. After a time the river washed away the bed and exposed the sewer. The sewer caused eddies in the river and these damaged the plaintiff's land. The council was held not liable in nuisance. Some damage is too remote. In *Lamb* v *Camden London Borough Council* [1981] 2 WLR 1038 (CA) it was held that it is not reasonably foreseeable that if in repairing a sewer an authority breaks a water pipe, causing a nearby house to be damaged and so left unoccupied, squatters will move into it and cause damage to the house. In such a case the authority is not liable in nuisance or negligence for the damage. In *Leonidis* v *Thames Water Authority* (1979) 77 LGR 722 it was held that loss of profit to a garage, caused when the nearby access road was obstructed for eleven months by repairs to a sewer, was recoverable from the water authority even though the work was done under statutory authority and did not take an unreasonably long time, the garage not being totally closed.

Although the authority is under a duty to provide such sewers as may be necessary to drain its area effectually it cannot be required to provide a sewer for every house wherever situated (*Kinson Pottery Co Ltd* v *Poole Corporation* [1899] 2 QB 41, at 49: house half a mile from existing sewer) unless the owner or occupier will give the water authority financial guarantees under s 16 of the 1973 Act. A planning application can be refused for inadequate sewerage arrangements despite the right under s 16 for the owner to demand a sewer, *Wimpey & Co Ltd* v *Sec of State and Maidstone DC* [1978] JPL 773.

A local authority must keep at its offices for public inspection a map showing all public sewers in its district and distinguishing those which are for surface water from those which are for foul water (1936 Act, s 32; 1973 Act, s 14(2)). The fact that a sewer is shown on such a map is not conclusive evidence that it has been duly adopted by the local authority by resolution under s 17 of the 1936 Act; however, it may have been acquired by the authority by some other method, such as by passing with the soil of a highway under the maxim 'quicquid plantatur solo, solo cedit' (*Royco Homes Ltd* v *Eatonwill Construction Ltd* [1979] Ch 276). Where an authority (not being a collection authority) provides pipes for collecting or disposing of waste or of anything produced from waste, or heat, air, steam or water produced from waste, or outfall pipes for sewage disposal, it must send a map showing the location of the pipes to the collection authority (Control of Pollution Act 1974, s 28). In England and Wales the collection authority is the district council or London borough (1974 Act, ss 12, 14 and 30).

Plans for a building or extension which are submitted for approval under the Building Regulations 1985 may be rejected or conditionally approved by the local authority if the building work is to be carried out over a sewer or drain (1936 Act, s 25), or pipe (1974 Act, s 28). For example, conditions may be imposed that the sewer is to be protected against settlement caused by the weight of the building. The authority may change the size of a sewer or the course it follows, but if this deprives a person of the use of it the authority must provide him with an adequate substitute and carry out any necessary alterations for him (1936 Act, s 22).

By s 26 of the Town and Country Planning Act 1971, and the Town and Country Planning General Development Order 1988 (SI No 1813) article 11(1)(d) and Sched 4, Part I, development consisting of the construction of buildings or other operations or the use of land for retaining, treating or disposing of sewage, trade waste or sludge (other than the laying of sewers, pumphouses in a line of sewers, septic tanks and cesspools serving single houses, single buildings or single caravans in which not more than ten persons normally reside or congregate) must be advertised in a local newspaper at least 21 days before being determined by the planning authority. In addition a notice in the prescribed form must be prominently displayed on the land for not less than seven days in the month immediately preceding the application. By

Sched 2, Part 16 of the 1988 Order any development not above ground on behalf of a water authority required in connection with the provision, improvement, maintenance or repair of a sewer, outfall pipe, sludge main or associated apparatus, has planning permission without the need of an express application. By Part 10 of the Order, works of inspection, repair or renewal of any sewer, main, pipe or other apparatus, including breaking open any land for that purpose, are also permitted without application for permission.

2 Sewer agreements

By s 18 of the Public Health Act 1936 as amended by s 14(2) of the Water Act 1973, a water authority may agree with any person who is constructing or proposing to construct a sewer that if the sewer is constructed in accordance with the terms of the agreement they will declare the sewer or works to be vested in them and the agreement will then become enforceable against the authority by the owner or occupier of any premises served by the sewer or works. In *Royco Homes Ltd* v *Eatonwill Construction Ltd* [1979] Ch 276, a local authority failed to pass a formal resolution under s 17 to adopt a sewer made by a developer but adopted as a public highway the lane through which it ran. It was held that the correspondence about adoption between the developer and the authority did not constitute an adoption agreement under s 18 but that the authority had 'acquired' the sewer under s 20 by virtue of the legal maxim that whatever is contained in the soil is part of the soil unless there is something to displace the effect of the maxim. (See the statement of the principle in *Simmons* v *Midford* [1969] Ch 415 at 420). No drain (as defined in s 34(1) of the Public Health Act 1936) can be a public sewer. It was held in *Hutton* v *Esher UDC* [1973] 2 WLR 917 (CA) that a pipe which had three functions, including that of a surface water sewer, could be a public sewer. It is common for authorities to impose their own criteria on the provision of pipes and pumping stations and to demand commuted sum payments for adoption under s 18 agreements. It is suggested that their discretion is wide but that they cannot demand more than sound sewer engineering practice would dictate, by analogy with *Royco Homes* v *Southern Water Authority* [1979] 1 WLR 1366 (HL).

3 Street works

The water authority has power to break open streets in order to lay such things as sewers and water pipes. The Act which confers the power will indicate the scope of it and the terms on which it is to be exercised, for example s 15 of the Public Health Act 1936, as amended by the Water Act 1973, Sched 8, para 32, but the procedure to be followed is laid down in the Public Utilities Street Works Act 1950, Pt II, ss 21 to 25. Where the highway authority (which may be the local authority) proposes to carry out works which will disturb the equipment of a statutory undertaker, such as a public gas supplier or Area Electricity Board, the authority must give notice to the undertaker specifying the works and the undertaker may serve notice in turn listing what it requires to be done to protect the apparatus from damage, at the authority's expense. A dispute will go to arbitration. Where it is the statutory undertaker which proposes to break up the street, Pt I of the 1950 Act lays down a similar procedure to the above for giving notice to the highway authority or, in the case of a private street, to those having the management or control of the street. By s 156 of the Highways Act 1980, the statutory undertaker may not, except in a case of emergency or with the consent of the highway authority, break up a street within 12 months of its having been reduced to less than two-thirds of its width by roadworks or having been resurfaced over one-third or more of its width.

Where a statutory undertaker causes a nuisance which is attributable to its carrying out of street works under its statutory powers, the authority is not liable unless it has also been negligent (*Department of Transport* v *North West Water Authority* [1983] 3 WLR 707 (HL), construing s 18(2) of the Public Utilities Street Works Act 1950). 'Negligence' in this context means failing to carry out the work and to conduct the operation with all reasonable regard and care for the interests of other persons, *Allen* v *Gulf Oil Refining Ltd* [1981] AC 1001, Lord Wilberforce at 1011.

4 Sewers in private land

The water authority may construct a public sewer through private land after giving reasonable notice to the owner and occupier (Public Health Act 1936, s 15, as amended by the Water Act 1973, Sched 8, para 32) even if it means demolishing a dwelling-house in order to lay the sewer (*Hutton* v *Esher Urban District Council* [1974] Ch 167 (CA)). 'Reasonable notice' is not defined in the Acts. A precedent of a notice is given in the *Encyclopaedia*, 4th ed, 16, 933. By s 90(4) of the 1936 Act a reference to a drain or sewer in Pt II of that Act (the Part includes s 15) includes a reference to manholes, ventilating shafts, pumps or other accessories. The power to construct a sewer does not permit the placing of a pumping-house on private land; to do that the authority must acquire the land by agreement or compulsorily under s 15 of the 1936 Act, and s 65 of the Water Resources Act 1963. The owner on whose land a sewer is to be constructed may not refuse but may claim compensation (1936 Act, s 278).

Where a public sewer is constructed in private land it is suggested, by analogy with private easements, that the right to construct carries with it the right to enter on the land subsequently to inspect, maintain and repair the sewer (*Jones* v *Pritchard* [1908] 1 Ch 630, at 638). Compensation is assessed on that basis (see Chapter 6/5). The sewer owner is, indeed, under a duty to repair (*Humphries* v *Cousins* (1877) 2 CPD 239). A water authority may construct sewers outside its district after giving notice in the press and to the authorities concerned (Water Act 1973, Sched 8, para 32). If there are objections a local enquiry must be held and the consent of the Secretary of State obtained to the construction (1936 Act, s 16; 1973 Act, s 1).

5 Compensation for sewer laying

Where the water authority places a sewer in private land compensation is payable for damage so caused (Public Health Act 1936, s 278). The authority acquires an interest in land (an easement) and damages are payable for (a) the easement and (b) the damages, present or future, arising from the acquisition, *Thurrock Joint Sewerage Board* v *Thames Land Co Ltd* (1925) 90 JP 1. The sum paid will include an element for possible future disturbance for maintaining and repair, a 'recognition payment' (*Markland* v *Cannock Rural District Council* (1973) 227 EG 1173). If damage is done on a future entry compensation will in addition be payable for that damage (*Markland's* case). If the work of laying a sewer is done negligently and subsidence occurs, an action is not statute barred six years after the work was done, under the Limitation Act 1980. Where such an excavation has been made there is a continuing nuisance and a fresh cause of action on each occasion of subsidence, *Crumbie* v *Wallsend Local Board*, [1891] 1 QB 503, *Smith Stone & Knight Ltd* v *Birmingham District Council* [1988] EGCS 49. It is no defence for the authority to plead that the works were done negligently and were therefore unauthorised, the authority is liable for the negligence of its sub-contractors, *Marriage* v *East Norfolk Rivers Catchment Board* [1950] 1 KB 284, *Smith Stone & Knight Ltd* (above).

The recent approach to assessment has been to consider four main factors, as follows:

(a) The trend of payments for pipelines of that type. An oil pipeline might be less desirable than a sewer, because of the possibility of greater damage from a leak.

(b) Payments for pipelines of that type in relation to capital land values. In *Markland's* case there was an award of 25p per linear yard for 1,247 yards for an 11-yard wide easement strip for a nine-inch sewer in farmland, with in addition 11 manholes at £5 each and two at £10 each, a total of £387.

(c) The trend of capital land values.

(d) The trend of payments for other types of pipelines.

These principles were enunciated in *Marklands's* case and were followed, in respect of a gas main, in *Wells-Kendrew* v *British Gas Corporation* (1974) 229 EG 272. In *Lucey's Personal Representatives and Wood* v *Harrogate Corporation* (1963) 14 P & CR 376, compensation of £75 for a wayleave 466 linear yards long for an 18-inch sewer pipe and six manholes in farm land was awarded. If the laying of the sewer has increased the value of the land by more than the loss which the laying has caused, no compensation is payable. The gain is set off against the loss, but the owner of the land does not seem to be liable to pay the water authority for any gain to him (1936 Act, s 278(4)). In *Rush & Tomkins Ltd* v *West Kent Sewerage Board* (1963) 14 P & CR 469, a compensation figure of £1,480 for laying a sewer was cancelled out by enhancement of value of part of the applicant's land by the sum of £3,800 because construction of the sewer made the land more readily saleable as industrial land.

In relation to compensation for pipe laying, matters to be considered include payments for a sewer easement the payment for which is in practice assessed on so much a metre of run, based on 50 per cent of the vacant possession value of a notional easement strip. In a helpful booklet, '*Sewers and Water Mains*', published by the Country Landowners Association (A5/1986) it is suggested that the minimum width needed to protect the pipe is ten feet on either side, plus the diameter of the pipe. In addition to payment for an easement there should be a claim for compensation for manholes, valves and other structures whether or not above ground. Under a national agreement made in 1986 British Telecom pay an annual wayleave of £11.47 for an above ground manhole, for loss to agricultural operations. The laying of the pipe will cause damage, disturbance and loss of profits and professional fees for negotiating all the matters may also be claimed. The British Standards publication, BS 8010: *Pipelines on Land: General*, should also be consulted.

6 Compulsory acquisition of land

Where the water authority finds it necessary to acquire land compulsorily under ss 15 and 306 of the Public Health Act 1936 and Control of Pollution Act 1974 Sched 3, para 27 for its operations in carrying out its duties of sewerage, it must follow the procedure laid down in the Acquisition of Land Act 1981 and the Compulsory Purchase Act 1965, giving due notice to all the persons with an interest in the land and making advertisements in a local newspaper. If there are objectors a public enquiry will be held and the order will require the confirmation of the confirming authority (1981 Act, s 13). In default of agreement the compensation for the land will be assessed by the Lands Tribunal. Compensation is also payable for the severance or injurious affection of other land not compulsorily acquired (Lands Clauses Consolidation Act 1845, s 49; Compulsory Purchase Act 1965, ss 7, 10). For example, in *Russell* v *Bradfield Rural District Council* (1957) 8 P & CR 432, a plot of land of 0.843 acre was acquired for a sewage works in the centre of a dairy farm of some 86 acres. The Lands Tribunal awarded the following items as compensation at 1957 values: for three unobtrusive manholes which did not interfere with cultivation, £4 each; for a sewer pipe line 313 yards long which might sterilise the land for building purposes for twelve feet on either side but was too deep to interfere with cultivation, £16; for an effluent pipe 209 feet long, £5; for cleansing a stream, £36; for fencing a ditch, £42; for severance and injurious affection of the farm because of the position and smell of the sewage disposal works, £400; even though it was argued that there are smells on farms anyway, the Tribunal took the view that people do not like sewage works on dairy farms and that their dislike would be reflected in the price that they would pay for the farm. The rules for assessing compensation in every case where land is compulsorily acquired or injuriously affected are laid down in the Land Compensation Act 1961, Pt II, ss 5–16, as supplemented by the Land Compensation Act 1973, ss 44–51.

7 Provision of sewer and connection of private premises

By s 16 of the Water Act 1973, the owner or occupier of premises may require the water authority to provide a public sewer to be used for domestic purposes for the drainage of the premises, provided that the owner has supplied a private sewer to be drained into it (for which he will need planning permission, *Wimpey & Co Ltd* v *Secretary of State and Maidstone District Council* (1978) JPL 773.) The meaning of 'public sewer' in this context apparently excludes one provided solely for the discharge of floodwater and rainwater (s 16(12)) and on one view this limits the meaning to a sewer for foul water. Some authorities however take the view that the section applies also to surface water sewers and accept a requisition for them, using s 15 of the Public Health Act 1936 (which entitles them to 'construct a public sewer' under a street and to acquire any sewer 'by agreement', 'sewer' being widely defined in s 20 of the Public Health Act).

In *William Leech (Midlands) Ltd* v *Severn Trent Water Authority* (1981) 260 EG 1123 (CA) a developer of a site of sixty-one houses brought the private sewer to point *A*. The water authority on receiving a requisition from him proposed to bring the public sewer to point *B* which was a higher level and in a different part of the site. A pump and rising main would have been needed to make the communication. It was held that the authority must provide a communicating sewer and not merely a sewer which the developer can cause to communicate, though the manner and place of communication are by s 16(3)(*a*) of the 1973 Act placed within the discretion of the water authority.

The owner or occupier who requisitions a public sewer must give certain financial guarantees to help towards the cost of the sewer, on a formula given in s 16(2) of the 1973 Act in the case of existing buildings and s 16(3) in the case of new buildings. In a case under s 16(3) the authority may require the deposit of a sum not exceeding the total cost of laying the sewer, as security for the annual payments, which payments may be taken out of the sum deposited (s 16(6), (7)). If the water authority is liable to provide a sewer under s 16 and fails avoidably to do so within six months or such longer period as the parties may agree, the authority is liable on summary conviction to a fine not exceeding £1000 (s 16(9)). This new power to requisition the provision of public sewers may help to avoid the situation where planning permission is refused on the ground that adequate sewers are not available to the proposed development.

A person is not liable for a full sewerage charge if his premises are not connected to a sewer (*Daymond* v *Plymouth City Council* [1976] AC 609) but can be liable to pay a reduced rate for such services as are provided by the authority (s 30 of the Water Act 1973, as amended by s 2 of the Water Charges Act 1976). In *South West Water Authority* v *Rumble's* [1985] 1 All ER 513 (HL) it was held that a person who had no water supply or drainage was nonetheless liable for a rate as the roof on premises over his shop drained into the sewer.

An owner or occupier of premises is entitled to have his drain or sewer connected to the public sewer. Before connecting he must give notice of his wishes to the water authority and the authority can refuse if the construction or condition of the drain is such that its connection would be prejudicial to the general system (1936 Act, s 34(3)). In Greater London the authority may refuse 'on such grounds as it thinks fit' (1973 Act, Sched 8, para 37(1)). If the authority consents, it may superintend the connecting up (1936 Act, s 34(4)) or it may insist on doing the work at the applicant's expense (s 36(1)). He may be required to lodge security for the expense (s 36(2)). No one may empty into the public sewer any matter which is likely to injure the sewer, interfere with its flow or prejudicially affect the treatment of its contents, or any chemicals or heated liquid if these things would be likely to be prejudicial to health, or any petroleum spirit or calcium carbide (s 27).

8 Waste

By s 1 of the Control of Pollution Act 1974 ('1974 Act'), it is the duty of each disposal authority (county councils and the Greater London Council in England, district councils in Wales: 1974 Act, s 30) to ensure that the arrangements for the disposal of waste are adequate to dispose of controlled waste. 'Waste' includes any scrap material or effluent or unwanted surplus substance arising from the application of any process and 'controlled waste' means household, industrial and commercial waste (s 30). A decision of a disposal authority that a person using plant to dispose of controlled waste within the meaning of the Control of Pollution Act 1974 required a disposal licence was not a decision amenable to judicial review, *R* v *London Waste Regulation Authority, ex parte Specialist Waste Management Ltd* (1988), *The Times* 1 November. Controlled waste cannot be deposited on land without a disposal licence on pain of criminal proceedings (s 3), and if the waste is poisonous or likely to give rise to an environmental hazard the penalties for unauthorised disposal are more severe (ss 3 and 5), see Waste Regulation and Disposal (Authorities) Order 1985 (SI No 1884). The Collection and Disposal of Waste Regulations 1988 (SI No 819) in force in part from 6 June 1988 and in part from 3 October 1988, replace the Control of Pollution (Licensing of Waste Disposal) Regulations 1976, as amended. The 1988 Waste Regulations are explained in Circular 13/88 (Welsh Circular 19/88). In *Ashcroft* v *Cambro Waste Products Ltd* [1981] 1 WLR 1349 the defendants had a licence to dispose of certain waste, under a condition that it must be covered to a certain depth. Oil waste and blue asbestos was deposited but was not duly covered. The breach was the responsibility of the site foreman; the respondent company and their site director did not know of it. It was held that to constitute an offence under s 3(1) it was not necessary that the respondents knowingly permitted the alleged breach of condition; it was enough that they knowingly permitted the deposit of controlled waste and that the deposit was in breach of the condition. The notification system introduced by the Deposit of Poisonous Waste Act 1972 has been abolished with the repeal of the Act (1974 Act, Sched 4). Radioactive waste may not be disposed of except under an authorisation granted by the Secretary of State under s 6 of the Radioactive Substances Act 1960, as amended by the 1974 Act, Sched 3, para 20. See Control of Pollution (Radioactive Waste) Regulations 1976 (SI No 959).

By s 6(2) of the 1974 Act a disposal licence for controlled waste may include, if the authority sees fit, '(b) the supervision by the holder of the licence of activities to which the licence relates; (c) the kinds and quantities of waste . . . the methods of dealing with them . . . (d) the precautions to be taken on any land to which the licence relates; (g) works to be carried out before the activities . . . are begun or while they are continuing; . . . a condition may require the carrying out of other works or the doing of any other thing which the authority considers appropriate'. It was held in *Attorney-General's Reference (No 2 of 1988)* (1989) *The Times*, 5 April (CA) that the powers under s 6(2) were not wide enough to allow a licence to contain conditions prohibiting public nuisances of all kinds whether or not they polluted water, endangered public health or constituted a serious detriment to the amenities of the neighbourhood. So a condition that 'the facility shall at all times be managed and operated so as to avoid creating a nuisance to the inhabitants of the neighbourhood', was not a proper condition. If as a result of the use of the plant a nuisance was *prima facie* committed there was no need for the authority to trace back to a particular failure of operation or management as there was *prima facie* an inference of failure; if the operator had taken all such steps as were reasonably open to him to ensure that a condition was complied with he had a defence under s 3(4)(*d*), (*Attorney-General's Reference* (above)).

The discharge of trade effluent into sewers may be permitted in certain circumstances, under the Public Health (Drainage of Trade Premises) Act 1937, and the Public Health Act 1961. Under these Acts as amended by the 1974 Act, no one may commence to drain trade effluent into a sewer without giving two months' notice to the disposal authority of the type and volume of the proposed discharge and obtaining the authority's consent (1937 Act, s 2(1); 1974 Act, ss 1, 30). Trade effluent is liquid, with or without particles in suspension, which is produced in the course of any trade or industry carried on at trade premises, excluding domestic sewage (1937 Act, s 14). In *Yorkshire Dyeing and Proofing Co Ltd* v *Middleton Borough Council* [1953] 1 WLR 393 it was held that where trade effluent was lawfully discharged from one set of premises and fresh premises were obtained in addition, the effluent from the new premises could not lawfully be sent into the sewer through the old premises without fresh notice and permission from the authority even though no greater quantity or different type of trade effluent was being passed into the sewer. It was held in *Thames Water Authority* v *Blue and White Launderettes Ltd* [1980] 1 WLR 700 (CA) that the discharge from launderette washing machines was trade effluent although it was the same as a discharge from domestic washing machines and 'domestic sewage' was excepted in s 14. It was held that in s 14 'domestic sewage' meant liquid from water closets and baths, lavatories and sanitary conveniences used by workpeople in a factory. Where trade effluent is discharged in breach of conditions contained in a water authority consent, a separate offence under s 32(1) of the 1974 Act is committed by each discharge and not by each condition broken, *Severn Trent Water Authority* v *Express Foods Group Ltd* (1988) *The Times*, 31 October.

The authority may give consent to drain trade effluent into a sewer upon terms which stipulate the volume, temperature, acidity or alkalinity and time of day of discharge: it may require that anything which would cause damage to the sewer or make treatment of the sewage more difficult must be extracted (Public Health Act 1961, s 59). In *Re Spenborough Urban District Council's Agreement* [1968] Ch 139 a company made an agreement under s 7 of the 1937 Act with the corporation to discharge effluent but no stipulation was made about ending the agreement. It was held on the construction of the document that either party could determine on reasonable notice and that the twelve months' notice given by the corporation was reasonable.

If an authority agrees to take waste at a fee which because of inflation becomes unrealistic the authority may be able to terminate the agreement on reasonable notice, if the court can construe the agreement as not having been intended to buy the discharging party 'perpetual immunity from progress and inflation' (Templeman J in *Beverley Corporation* v *Richard Hodgson & Sons Ltd* (1972) 225 EG 799; similarly *Re Spenborough Urban District Council's Agreement* [1968] Ch 139, and *Staffordshire Area Health Authority* v *South Staffordshire Waterworks Co* [1978] 1 WLR 1387 (CA).

A person who owns or occupies land on or under which is any sewer or drain used or intended to be used for discharging trade effluent into an authority's sewer must, on the authority's written request, furnish to it all the information and plans he has concerning the sewer (1937 Act, s 9). The disposal authority has power to enter premises and take samples of trade effluent passing from them into the local authority's sewers (Public Health Act 1936, s 287; Public Health (Drainage of Trade Premises) Act 1937, s 10).

Guidance on the principles to be applied in permitting the discharge of effluent was given by the Minister of Housing and Local Government in circulars on trade and sewage effluents, Nos 37 of 1966 and 64 of 1967.

Under ss 4 and 7 of the 1937 Act discharges made or permitted before that Act commenced were not subject to the Act. However, these discharges are brought under the control of the 1937 Act by ss 44 and 45 of the 1974 Act.

By s 31(1) of the 1974 Act it is an offence to cause or knowingly permit, without a licence or consent of the Secretary of State or water authority, any poisonous, noxious or polluting matter to enter any stream or coastal waters or specified underground water ('relevant waters' as defined in s 31(1)(*a*)). A person can be liable under s 31(1)(*a*) of the 1974 Act without proof of intent to pollute or negligence unless the operative cause of the escape was an intervening event outside his control or an Act of God. If he conducts some active operation involving the storage or use of materials which, if they escaped, would cause pollution he may thus be liable, *Southern Water Authority* v *Pegram*, (1989) *The Times*, 1 March. In this case an ingress of rainwater (not unusually heavy) which had flowed into a stream indirectly from a blocked drain was not such an intervening act and the owner was liable. Action in an emergency to avoid danger to the public provides a defence (s 31(2)). By s 32(1) it is an offence to cause or knowingly permit (*a*) any trade effluent or sewage effluent to be discharged into relevant waters or from land in Great Britain through a pipe into the sea outside coastal waters or from a building or plant onto land or into any lake, loch or pond which does not discharge into a stream; (*b*) any matter other than trade or sewage effluent to be discharged into relevant waters from a sewer or from a drain of a highway authority. It is a defence if a water authority's discharge into a sewer is attributable to a discharge made by another person and the authority could not reasonably have been expected to prevent it (s 32(5)). A person can be liable for 'causing' polluting matter to enter water even though it escapes without negligence on his part (*Alphacell Ltd* v *Woodward* [1972] AC 824 (HL) (decided under s 2 of the Rivers (Prevention of Pollution) Act 1951), followed in *Price* v *Cromack* [1975] 1 All ER 113). In *FJH Wrothwell Ltd* v *Yorkshire Water Authority* (1983) *The Times*, 31 October, a director of the defendant company deliberately poured into its drains 12 gallons of a concentrated herbicide known to be toxic to fish, meaning it to discharge in the public sewer. By a system of drainage unknown to him it entered a stream and not the sewer. The company was liable for 'causing' it to flow into the river. The word 'cause' was a simple word in everyday use and must be construed by the exercise of common sense applied to the facts of the case. The company was liable under s 2 of the Rivers (Prevention of Pollution) Act 1951 and s 4 of the Salmon and Fresh Water Fisheries Act 1975, but not under s 27 of the Public Health Act 1936. If a third person interferes with plant or equipment and thereby effluent enters a stream, the owner of the plant is not liable for 'causing' it to do so (*Impress (Worcester) Ltd* v *Rees* [1971] 2 All ER 357).

9 New buildings

The local authority (the district council or London borough council, Building Act 1984, s 126) may not approve plans for a new building unless it is satisfied that satisfactory provision is made for drainage or that such provision may be dispensed with (Building Act 1984, s 21). Unless a proposed drain connects with a sewer or cesspool, it cannot be considered satisfactory (s 21(4), 1984 Act). But if the plans show that the drainage for the building will be satisfactory, consent to the application cannot be refused on the ground that the sewer to which the drain will connect is unsatisfactory (*Chesterton Rural District Council* v *Ralph Thompson Ltd* [1947] KB 300). If there is not a sewer within 100 feet of the building, connection to it cannot be insisted upon unless the local authority agrees to bear the cost of making that part of the connection which is beyond 100 feet (s 21(4),(5)). A formal resolution of the authority is necessary to make the authority liable; the agreement of their surveyor alone was held not binding on the local authority in *Princes Investments Ltd* v *Frimley and Camberley Urban District Council* [1962] 1 QB 681. In *Myrick* v *Pembroke Corporation* (1912) 76 JP 365, a house was over 119 feet from the available sewer but there was a privy at the boundary wall of the house less than 30 feet from the sewer and an outbuilding less than 99 feet from the sewer. When the owner failed to comply with a notice to connect, the local authority coverted the privy into a water closet and successfully claimed the cost from the owner as it was held that the building with within 100 feet of the sewer because its outbuildings were within this distance. Connection with the sewer cannot be insisted on if the sewer is not at a level which makes it reasonably practicable to connect to it, or the intervening land is not land through which the owner is entitled to put his drain, or the sewer is not a public one or one that he is entitled to use (s 21(4)). The local authority must be satisfied that there is satisfactory provision for the drainage of both waste water and rainwater from the proposed new building (s 21(2)). See also Building Regulations 1985 (SI No 1065) discussed at para 13 (later).

10 Drainage of existing buildings

The local authority has powers concerning the drainage of existing buildings: where satisfactory provision has not been made for drainage and ought to be made, or any cesspool, private sewer or drain is insufficient or admits subsoil water or is in a condition making it prejudicial to health or a nuisance by reason of being no longer used, or for any other reason, then the local authority can give notice requiring the owner or occupier to carry out the necessary works to remedy these defects (Building Act 1984, s 59(1)). If premises, though adequately drained, are served by a system which is not adapted to the general sewerage scheme of the neighbourhood or is otherwise objectionable, the water authority may, at the authority's expenses, insist on the premises being connected to and using the general sewerage scheme in place of the former system (s 42(1) and Water Act 1973 s 14(2)). If the drainage is ineffectual the local authority can use the provisions of s 39 and require necessary works to be done at the owner's or occupier's expense.

A pipe used to convey rainwater from a roof may not be used also as a soil pipe from a sanitary convenience and a pipe for surface water may not act as a ventilating shaft for a foul water pipe (Building Act 1984 s 60(1)). The soil pipe from a water-closet must be properly ventilated (s 60(2)). If the contents of a cesspool overflow or leak away the local or water authority may by notice require the owner or occupier to carry out such works as are necessary to prevent this from recurring (s 50; Water Act 1973, Sched 8, para 40). If it appears to the local authority that a drain, private sewer, water-closet, waste pipe or soil pipe is not sufficiently maintained and kept in good repair and that the necessary work will not cost more than £250, it can serve notice requiring the work to be done, and if it is not done within seven days it can carry out the work itself and charge the owner or occupier with the cost (Public Health Act 1961, s 17(1)). In an urgent case, where it appears to the local authority that a drain, private sewer, water-closet or soil pipe is stopped up, the authority may give notice in writing for the blockage to be cleared within forty-eight hours, and will then carry out the work at the expense of the owner or occupier if the notice is not complied with (Public Health Act 1961, s 17(3)).

A person in a local authority area may not alter the course of any underground drain which communicates with a sewer or cesspool, except in an emergency, without giving at least 24 hours notice to the local authority of his intention to do so; the local authority must be given free access to the work (Building Act 1984, s 61).

If the local authority or water authority considers that it has reasonable grounds for believing that a drain, cesspool, closet or private sewer is prejudicial to health or a nuisance, or if the water authority believes on reasonable grounds that a drain or private sewer is admitting subsoil water into a public sewer, it may test the suspected thing and if necessary open up the ground (1936 Act, s 48(1); 1973 Act, Sched 8, para 39). If the suspicion proves justified, the appropriate notices to rectify will be served; if unjustified, then the local or water authority must reinstate and pay compensation for any damage caused in the investigation (s 48(2); para 39) By s 48(1)(A) of the 1984 Act the relevant authority is the water authority if there is a connection to a public sewer, and in other cases the local authority.

11 Sanitary conveniences

The local authority must ensure the proper provision and maintenance of sanitary conveniences in buildings. A local authority may not pass plans for a new building or extension of an existing building unless sufficient sanitary accommodation is provided, Building Regulations 1985 Sched 1, G4 and H Plans deposited must show satisfactory provision for drainage, s 21(1) 1984 Act. A sewer within 100 feet of the building is available if it is at a reasonably practicable level for drainage into it and the owner or occupier is entitled to drain through the intervening land to it (s 21(4)). The local authority may insist on a connection to a sewer more than 100 feet away if the authority bears the extra cost of the connection (s 21(5)). Where the new premises are to be used as factories (defined in s 175 of the Factories Act 1961) or as offices, shops or railway premises there must be satisfactory separate closet accommodation for each sex (1961 Act, s 9; Offices, Shops and Railway Premises Act 1963, s 7).

Where an existing building appears to the local authority to have insufficient closet accommodation, or any closet is prejudicial to health or a nuisance and requires reconstruction to reach a satisfactory condition, the owner may be given notice to provide necessary closets (Building Act 1984, s 64). The closet must be a water closet if there is a sufficient water supply and sewer available within 100 feet, and the other provisions with regard to sanitary conveniences in new buildings outlined above will apply (s 64(2)). If the existing closet is in a state prejudicial to health or is a nuisance but can be put into satisfactory condition without reconstruction, a notice requiring this to be done can be served (Public Health Act 1936, s 45). A reasonable number of sanitary conveniences must be provided by the owner or occupier of any betting shop or any place used or proposed to be used for entertainment, exhibition, or sport to which members of the public are admitted either as spectators or otherwise, or sale of food or drink to members of the public for consumption there (Local Government (Miscellaneous Provisions) Act 1976, s 20); and the needs of disabled persons must be borne in mind, so far as is practicable and reasonable (Chronically Sick and Disabled Persons Act 1970, ss 5 and 6. See *Code of Practice for Access of Disabled to Buildings* BS 5810: 1979).

A county council or other local authority is empowered to erect public sanitary conveniences in proper and convenient situations (1936 Act, s 87; Local Government Act 1972, Sched 14, para 1). It must bear in mind the needs of disabled persons (1970 Act, above, s 6). If the building is not by the county council and if the convenience is on or under a highway the consent of the county council, as the highway authority, is needed (1936 Act, s 87; 1972 Act, Sched 21, para 1). No one else may provide a public convenience (except on a railway station) without the consent of the local authority (1984 Act, s 68). The local authority may supply on loan temporary sanitary conveniences if those at any premises need to be disconnected, for repair, for example, 1984 Act, s 67.

12 Working of minerals

There are restrictions on the right of an owner of land to work minerals under his land if by so doing he might interfere with the 'sanitary works' (sewers, drains, water pipes and similar things) of the water authority (Public Health Act 1875 (Support of Sewers) Amendment Act 1883, and the new s 20 of the Public Health Act 1936, which was supplied by the Water Act 1973, Sched 8, para 33). The general position is that a landowner may work his minerals in a reasonable and proper manner without being liable for any damage so caused to the sanitary works but that he is liable for damage otherwise caused (1883 Act, s 4). The authority may require him not to work under or near to the sanitary works but in that case must compensate him for the value of the minerals which he has to leave in the ground and for the additional expense of mining in this way. The owner who proposes to work minerals within forty yards laterally of the reservoirs, buildings, pipes or other conduits or underground works of a water authority or at any distance underneath them must give to the water authority thirty days' notice of his intention and they may prevent the work if they apprehend damage to such works and if they agree to pay compensation (Water Act 1945, Sched 3, paras 13 and 14).

13 Building Regulations 1985

Local authority building byelaws have been superseded by building regulations made under the Building Act 1984, s 1. The current regulations are the Building Regulations 1985 (SI No 1065), which replace the 1976 and 1981 Regulations. The Building Regulations apply to the construction of new buildings in England or Wales and the alteration or extension of existing buildings. Notification to the local authorities in whose area the work will be carried out is necessary. These are the local district council or outer London borough council. In Inner London boroughs the relevant provisions are the Building (Inner London) Regulations 1985 (SI No 1936) and 1987 (SI No 798) which apply the Building Regulations 1985 with small modifications.

The 1985 Regulations differ markedly from their predecessors in that they impose less detailed control but by s 6 the Secretary of State is empowered to issue practical guidance documents which supply the detail which was formerly in the Regulations. He has issued the comprehensive Building Regulations 1985 Manual (Sweet and Maxwell's *Encyclopaedia of Environmental Health*, vol 3, p 4775).

The local authority which enforces regulations owes a duty of care to successive owners of the premises (subject to a limitation period, Latent Damage Act 1986, amending the Limitation Act 1980) which it has inspected and passed to see that the regulations have been complied with, *Dutton* v *Bognor Regis UDC* [1972] 1 QB 373 (CA); *Anns* v *Merton London Borough Council* [1972] 2 WLR 1074 (HL). A duty can be owed even though the owner also employs a builder and an architect to supervise the work, *Acecrest* v *Hattrell and Partners* [1982] 3 WLR 1076. A person who is having building work done may at his option choose to have it inspected by an 'approved inspector' and in such a case the local authority's functions of inspection will not be exerciseable, ss 47–53. By s 54 approved public bodies may be their own inspectors, on terms (Sched 4). Both provisions are covered by the Building (Approved Inspectors) Regulations 1985 (SI No 1066).

By reg 11 a person who intends to carry out building work or to make a material change of use shall give to the local authority a 'building notice' in accord with reg 12 or deposit plans with the authority under reg 13. Consultation with the fire authority may also be necessary, reg 11(2).

Before building work is commenced and at specified stages of the work, notice must be given to the authority (at least 48 hours before commencing work and at least 24 hours before covering any drain or private sewer, reg 14. The local authority may test any such drain or private sewer to establish whether it complies with Part H of Sched 1, reg 15.

By Sched 1, concerning:

(i) Site preparation Precautions shall be taken to avoid danger to health from substances on or in the ground, C2. Subsoil drainage shall be provided to avoid ground moisture passing into the building, C3.

(ii) Means of ventilation An adequate supply of air shall be provided for persons in the building, F1.

(iii) Hygiene A bathroom containing a fixed bath or shower bath shall be provided for all dwellings, G4. Sufficient sanitary conveniences shall be provided, G4.

(iv) Sanitary pipework and drainage Any system carrying foul water from the premises to the foul water outfall shall be adequate, H1. Cesspools, septic tanks and settlement tanks shall be constructed to permit access for emptying and to avoid contamination by their contents, H2. The rainwater drainage system shall be adequate, H3.

(v) Heat producing appliances Fixed heat producing appliances must have an adequate supply of air for efficient combustion and adequate means of discharge of the products of combustion. They must be so placed as to reduce the risk of causing fire, J1 to 3.

(vi) Insulation of heating services Heat water pipes and warm air ducts shall have adequate thermal insulation unless they are intended to contribute to the heating of the building, L5.

The Building Regulations Manual gives detailed guidance and illustrations on all the above matters.

By Sched 3 buildings controlled under other legislation, (such as the Nuclear Installations Act 1965 or the Ancient Monuments and Archaeological Areas Act 1979); buildings not frequented by people and greenhouses and agricultural buildings, temporary buildings and some others are exempt from the Building Regulations.

14 Statutory nuisances

It is the duty of a local authority to inspect its district from time to time to detect any statutory nuisance on premises and to serve a notice on the person causing or permitting a nuisance, or, if he cannot be found, then on the owner or occupier of the premises, requiring him to abate it and to carry out such work as shall be necessary for that purpose (Public Health Act 1936, ss 91–93). If a notice is not complied with, or in the opinion of the local authority the nuisance is likely to recur, it may apply to the magistrates' court for a nuisance order requiring the defendant to comply or to carry out any work necessary to prevent a recurrence, or both (s 94). If the occupier has not caused the nuisance and the person responsible cannot be found the local authority may abate the nuisance and charge the cost to the person liable (s 93). Examples of statutory nuisances are any pond, pool, ditch, gutter or watercourse which is so foul or in such a state as to be prejudicial to health or a nuisance (ss 92, 259; a sewer is not included in this phrase: *R* v *Parlby* (1889) 22 QBD 520), and any well, tank, cistern or water-butt used for the supply of water for domestic purposes which is so placed, constructed or kept as to render the water liable to be contaminated to an extent prejudicial to health (ss 92, 141). There is also power to close a polluted well, tank or other source of supply (s 140). Nuisances which are likely to recur can be dealt with by a prohibition notice served by the local authority under the Public Health (Recurring Nuisances) Act 1969, s 1(1) and a magistrates court can make a nuisance order under s 2(2).

Chapter 9

Sewers and Drains After Privatisation

The Water Act 1989 transfers to sewerage undertakers the functions of water authorities relating to the provision of sewerage services and Sched 8 of the Act makes amendments to the enactments relating to the transferred functions (s 68). One company can be both the water undertaker and the sewerage undertaker for the area.

By s 66 it is the duty of every sewerage undertaker:

(a) to provide, improve and extend such a system of public sewers (whether inside its area or elsewhere) and so to cleanse and maintain those sewers as to ensure that the area is and continues to be effectually drained: and
(b) to make provision for the emptying of those sewers and such further provision (whether inside its area or elsewhere) as is necessary from time to time for effectually dealing, by means of sewage disposal works or otherwise, with the contents of those sewers (s 66(1)).

The duties of the undertaker shall be enforceable by the Secretary of State under s 19 or by the Director General of Water Services. The Secretary may make regulations which will facilitate the determination of the extent to which an undertaker is failing in its duties and the regulations can include provisions for payment of money by an undertaker to a person who is affected by a failure to reach performance standards (s 67(2) and (3).

1 Requisitioning of sewers

It shall be the duty of a sewerage undertaker to provide a public sewer to be used for the drainage for domestic purposes of premises in a locality if a person entitled to do so requires it by notice (s 69(1)). The persons so entitled are the owner or occupier of premises, the local authority, a Commission for New Towns, the Development Board for Rural Wales or an urban development corporation (s 69(3)). The drainage of domestic premises means the removal of water used for cooking or washing (not being from a laundry or food preparers), water from lavatories and surface water (s 69(2)). The undertaker shall not be in default unless it has not laid a public sewer within six months of the 'relevant day', (s 69(4)) but that day is not until any reasonable financial stipulations imposed by the undertaker as a condition of laying the sewer have been agreed to by the requisitioner (ss 69(9) and 70). If the drainage charges are not likely to meet the sewer costs the undertaker may impose a requirement that the deficit is met by the requirer for up to 12 years (s 70). A dispute on the undertaking required may be referred to an arbitrator or in default of agreement to the president of the Institution of Civil Engineers (s 70(8)). A local authority may by agreement with the undertaker carry out such part of the undertaker's sewerage functions in the area as is agreed but the undertaker will remain liable for performance of the functions (s 71).

2 Controlling and remedying pollution

Chapter I of Part III of the Act (ss 99 to 111 inclusive) applies to 'controlled waters', which are defined in s 99. By ss 100 and 101 the Secretary of State may prescribe water quality objectives in respect of those waters and may specify to the National Rivers Authority the acceptable level of water quality for waters of particular descriptions; the Authority is under a duty to achieve and maintain those standards and to monitor the extent of pollution in the controlled waters (s 102). Controlled waters are: relevant territorial waters, coastal waters extending landwards from the baselines from which the three-mile territorial limit is measured; inland waters of any lake, pond or river above the fresh-water limit; and ground waters in underground strata (s 99(1)).

A person commits an offence if he causes or knowingly permits any poisonous, noxious or polluting matter or any solid waste matter to enter any controlled waters; or matter, other than trade effluent or sewage effluent, to enter controlled waters by being discharged from a pipe, drain or sewer in contravention of a prohibition; or any trade effluent or sewage effluent to be discharged into any controlled waters or from a pipe into the sea outside the limit of controlled waters; or into any waters of a lake or pond; or permits any matter to enter into inland waters so as to tend to impede the flow of the waters so as to lead or be likely to lead to a substantial aggravation of pollution due to other causes or to the consequences of such pollution (s 103(1)).

It is a defence if a person has a disposal licence to allow his effluent into the waters, or a consent under the 1989 Act or the Control of Pollution Act 1974 (s 104(1)). A sewerage undertaker is not guilty of an offence under the Act if the contravention is attributable to a discharge which another person has made into the undertaker's sewer or works and the undertaker was not bound to receive it there and could not reasonably have been expected to prevent it (s 104(7)). Regulations may be made by the Secretary of State which prevent a person from having custody or control of any poisonous, noxious or polluting matter unless prescribed precautions have been made for preventing it from entering controlled waters (s 106(1)); byelaws may be made by the Authority for prohibiting the washing or cleaning of anything prescribed and for controlling on waters vessels which are provided with water closets or other sanitary appliances (s 109(1)). The Authority may in certain cases carry out anti-pollution works of prevention and control where there is or is likely to be an unauthorised discharge into controlled waters and may charge the expenses to the person who caused or knowingly permitted the discharge (s 110). The Secretary of State and the Minister jointly may prescribe codes of good agricultural practice for avoiding or minimising the pollution of controlled waters (s 111). It is the duty of the Authority and water undertakers to exchange information concerning pollution incidents (s 114).

The Authority or the water undertaker may construct and maintain on land belonging to them drains, sewers, watercourses, catchpits and other works to intercept, treat and dispose of any foul water arising or flowing upon that land or of otherwise preventing the pollution of waters whether surface or underground from which they are entitled to take water (s 148).

3 Compulsory acquisition

The National Rivers Authority and a water or sewerage undertaker may be authorised by the Secretary of State (or a Minister in the case of the Authority) to purchase compulsorily any land anywhere in England or Wales which is required by the Authority or the undertaker for its purposes in carrying out its functions (s 146(1)). The power includes creating and then acquiring new rights and also extinguishing rights (s 146(2)). By s 147 pipes and sewers laid by the Authority or an undertaker under the provisions of Sched 18 shall vest in the body which laid them. By s 154 nothing in the Act which empowers a body to carry on works shall entitle them to affect injuriously (unless with consent) works or property of certain undertakings listed in s 154(3), being undertakings of the Civil Aviation Authority, the British Coal Corporation, an undertaking running a telecommunications system, a public gas supplier, the CEGB or Area boards (or their successors), a public utility undertaking, British Railways Board or London Regional Transport. The procedure for compulsory acquisition is contained in Sched 19. The Director may be called upon to investigate a complaint made by an owner of private land in respect of the use by the Authority or a water or sewerage undertaker of their powers; he may direct a sum not exceeding £1,000 to be paid in respect of the owner's loss, damage or inconvenience (s 155).

4 Repeals

The Water Act 1989 does not (unlike the Electricity Act 1989) repeal substantially the whole body of legislation on which the previous position rested. The main statutes, such as the Public Health Act 1936, the Public Health (Drainage of Trade Premises) Act 1937, the Water Resources Act 1963, the Control of Pollution Act 1974, and the Land Drainage Act 1976 remain, with appropriate amendments where, for example, 'water authority area' is referred to. A number of sections of the Water Acts 1945 and 1973 are repealed and replaced. The following Acts are repealed entirely: The Drought Act 1976, The Water Resources Act 1971, The Border Rivers (Prevention of Pollution) Act 1951.

Chapter 10

Land Drainage Before Privatisation

The responsibility for land drainage was from 1 April 1974 placed upon nine regional water authorities in England (and the Greater London Council and the London borough councils) with the Welsh National Water Development Authority acting in Wales, by the Water Act 1973, and upon the internal drainage boards set up under the Land Drainage Act 1930. Part of the 1973 Act and the whole of the 1930 Act were repealed by the Land Drainage Act 1976 ('1976 Act') which, however, continued the existing arrangements (1976 Act, ss 1 and 6, Scheds 1 and 2).

Ministerial authority for land drainage rests with the Minister of Agriculture, Fisheries and Food (1976 Act, s 116(1)).

Each water authority must arrange for the discharge of its drainage functions (except those of levying charges and borrowing money) through a regional land drainage committee (1976 Act, s 1(1)). This committee establishes a local land drainage committee for each district (1976 Act, ss 4 and 5).

1 Powers of water authorities

The drainage authority of each area has powers to cleanse, repair or otherwise maintain in an efficient state any watercourse or drainage work; to improve any existing watercourses or drainage work, such as by widening, deepening, straightening, removing or altering any obstruction such as a dam or weir; and to construct new works such as watercourses required for the drainage of the area (1976 Act, s 17).

The term 'watercourses' includes all rivers, streams, ditches, drains, cuts, culverts, dykes, sluices, sewers and passages through which water flows (1976 Act, s 116(1)).

An authority may (on giving reasonable notice) take up or

divert the drains, culverts or pipes of any local authority if they interfere with the improvement or alteration of any watercourse, provided that the work is done at the authority's expense and that other drains, culverts and pipes equally effective are substituted (Public Health Act 1936, s 330).

Where in the opinion of the authority the proper flow of water in a watercourse is impeded, the drainage authority may serve notice on the person in occupation or control of the land adjoining the obstruction or the person responsible for causing it, requiring him to remedy it (1976 Act, s 18). An appeal against the notice can be made to the magistrates within 21 days of the service of the notice (s 19).

A person may not obstruct the flow of any watercourse such as by a dam, weir or culvert without the permission (not to be unreasonably withheld) of the drainage authority, or alter any obstruction in a watercourse without permission (1976 Act, s 28). If he does so he is deemed to have created a nuisance which he may by notice be ordered to abate within a specified time (ibid, subs (4)).

A person may not without the consent of the water authority erect any structure in, over or under a watercourse which is part of a main river or alter or repair any such structure if the work is likely to impede the flow of water or impede any drainage work (1976 Act, s 29).

A drainage authority can deposit on the bank of a watercourse or on land adjoining it any matter removed from the watercourse in the process of dredging, widening or deepening it, unless a statutory nuisance within the meaning of the Public Health Act 1936, Pt III, would be caused by so doing (1976 Act, s 33). Where a person suffers damage by such depositing, the authority may compensate him if it thinks fit, but where he suffers damage which the authority could by taking reasonable care have avoided doing it must compensate him (ibid). He cannot claim damages, as distinct from compensation, unless he can show negligence in the manner of carrying out the work. In *Marriage* v *East Norfolk Rivers Catchment Board* [1950] 1 KB 284, the (then) board dredged a river and deposited the dredgings on one bank. This prevented flood water from escaping over the bank, so the water went into another channel in abnormal quantity and carried away the plaintiff's bridge. He failed in a claim based on nuisance and negligence to show negligence in the manner of carrying out the work, and was therefore limited to the compensation provided for by statute. The powers of the authorities are permissive. If they fail to exercise their powers at all or exercise them honestly but ineffectively, they are not liable for damage sustained by reason of the failure or partial failure to exercise (*East Suffolk Rivers Catchment Board* v *Kent* [1941] AC 74). If the authority is under a duty to act, as distinct from having merely a power to act, it is a question of construction whether an individual affected has a remedy in tort. For example, the provision in the statute of a remedy may exclude any action in tort. In *Hesketh* v *Birmingham Corporation* [1924] 1 KB 260 the corporation owned a sewer which ran alongside a natural stream and at times of heavy rain outlets relieved pressure on the sewer by discharge into the stream. Because of building on neighbouring land, in the course of time the outlets became inadequate and after a heavy storm the land became flooded. Held: No action would lie for negligence for failure to enlarge the capacity of the stream, the remedy lay under s 299 of the Public Health Act 1875 with assessment of compensation by arbitration under s 308.

A navigation, harbour or conservancy authority may divert or alter the level of any watercourse under the control of a drainage authority if the watercourse interferes with the alteration or improvement of any river, canal, dock, harbour, basin or other work which is under the control of the former. The altered or substituted watercourse must be equally effective, though (1976 Act, s 106).

2 Compensation for damage

Where a person suffers damage or loss because of the exercise by a drainage authority of its general powers he can (except where otherwise provided) claim compensation, to be assessed by the Lands Tribunal if agreement cannot be reached (1976 Act, s 17(5)). In *Glazebrook* v *Gwynedd River Board* (1964) 15 P & CR 75, the board carried out drainage works in a stream running through the claimant's dairy farm. He alleged that this reduced the amount of drinking water for his cattle, but as it was held on the facts that the benefit to his land by the drainage exceeded any loss no compensation was awarded. In *Welsh National Water Development Authority* v *Burgess* (1974) 28 P & CR 378 (CA), a person with fishing rights on a river was awarded £700 damages for diminution of the value of the fishing when the water authority's predecessor (the Gwynedd River Authority) 'retrained' the river, so as to avoid further flooding after two heavy floods had caused substantial deposits of gravel in the river. The works, though necessary, were too drastic and paid no adequate regard to the fishing interests.

3 Compulsory acquisition

An internal drainage board may acquire land compulsorily for its statutory purposes, and by agreement (1976 Act, s 37(1)). It may secure the creation of new interests in or over land (1976 Act, s 37(3)). It is likely that the term 'new interests or rights' is wider than an interest with the characteristics of a common-law easement; a statutory right derives its validity from statute and not from the common-law principles. There may, for example, be no dominant tenement which the drainage right benefits.

4 Byelaws

Drainage authorities have power to make such byelaws as they consider necessary to secure the efficient working of the drainage system in the area (1976 Act, s 34). They may in particular prevent the discharge into the watercourse of matter likely to obstruct it, and prevent the improper use of the watercourse or its banks to preserve it from damage (1976 Act, s 34(2) and Sched 4). Byelaws made under the power do not take effect until confirmed by the Minister of Agriculture (Sched 4, para 1).

5 Powers of a council

By s 97 of the 1976 Act, a county council, district or London borough council or the Common Council may exercise the powers conferred on internal drainage boards by s 18 of the 1976 Act (to maintain the flow of a watercourse). These councils may, by s 100, have the powers conferred by s 17 (general drainage powers) and s 33 (widening and dredging watercourses) if it would be impracticable to constitute an internal drainage board, such as where the area to be drained is quite small.

6 Drainage of another's land

Where a ditch (whether open, culverted or piped) (1976 Act, s 40(4)) is in such a condition as to cause injury to land or to prevent the improvement of drainage, the Agricultural Land Tribunal may order any person named in the order to carry out such work for putting it into proper condition as may be stated in the order (s 40(1)). If within three months the work is not done the Minister of Agriculture or any drainage board authorised by the Minister may enter on the land, do the work and recover from the person named any expenses so incurred (s 43(1), (2)). There are similar provisions where the drainage of land requires work on a ditch on adjoining land (s 41). The Agricultural Land Tribunal may authorise an applicant to enter the adjoining land to carry out the specified work. By s 44 where any person sustains injury by reason of the exercise of any power conferred by ss 40 to 44 (Pt III of the Act) he is entitled to compensation, to be assessed by the Lands Tribunal in a disputed case, unless the power was exercised for the purpose of executing works which the person was required to do by an order made under Pt III.

Chapter 11

Land Drainage After Privatisation

The Water Act 1989 dissolves the existing water authorities which had the function of land drainage under the Land Drainage Act 1976 and transfers the authorities' property, rights and liabilities under schemes made under Sched 2, which divide them between their successor private companies and the National Rivers Authority (s 3).

The Land Drainage Act remains effective, with necessary amendments as shown in Sched 24 of the 1989 Act (though ss 1 to 3 are repealed). The private companies take on both the functions of water supply undertakers and of sewerage undertakers though in any particular area they may not have all the functions throughout their area. For example, there could be existing water supply companies which remain the water undertakers by appointment of the Secretary of State under s 3 for a particular part of the area.

By s 131 and Sched 14 the flood defence duties of the former water authorities are transferred to the National Rivers Authority and will be carried out by flood defence committees, regional and local, set up under ss 132 to 134.

For the purpose of the drainage of land the internal drainage districts within the areas of the former water authorities will act as internal drainage districts within the areas of the regional flood defence committees; and for each district there continues, in accordance with s 6 of the Land Drainage Act 1976, to be a board known as an internal drainage board, which is a body corporate (s 135).

The compulsory powers of a water and sewerage undertaker to purchase any land which is required by it for the purposes of or in connection with the discharge of its functions (including the creation and acquisition of new rights) are contained in s 146(1) and (2) and the procedure to be followed is that under the Acquisition of Land Act 1981 (s 146(4)).

Chapter 12

Electricity Before Privatisation

The generation of electricity throughout England and Wales is the responsibility of the Central Electricity Generating Board. The Board must also supply electricity in bulk to the twelve area boards for distribution to consumers (Electricity Act 1957, s 2). The Board and area boards had a monopoly of the public supply and distribution of electricity (Electricity Act 1947, s 57) but that was removed from 1 June 1983 by s 1 of the Energy Act 1983 (the whole Act came into force on 1 September 1983).

1 Duty to lay street mains

An area board must lay down on being requisitioned to do so suitable and sufficient distributing mains for a general supply of energy for ordinary consumers (including a supply to public lamps unless otherwise agreed with the local authority) throughout every street, part of a street or other route in its area (Electric Lighting (Clauses) Act 1899, Sched, s 21). (The Sched to the 1899 Act was incorporated with the Electricity Act 1947 by s 57(2) and Sched 4, Part III of the 1947 Act.) Any six or more owners or occupiers of premises along a street, part of a street or route may serve written requisition on the board requiring it to lay down a distributing main for a general supply (Sched, s 24); a local authority which has the control or management of street lamps may also requisition (Sched, s 29). Where the requisition is made by owners or occupiers the board may within fourteen days serve a counter-notice on them declining to supply unless they undertake to take for three years at least such an amount of energy as the counter-notice specifies, calculated in a manner laid down in the Act (Sched, s 25). If the board does not serve a counter-notice it is bound to carry out the requisition. If a local authority requisitions, it must within fourteen days of its requisition tender an agreement to the board to take a supply to the public lamps for at least three years (Sched, s 26).

2 Obligation to furnish a supply

An owner or occupier of any premises which are within fifty yards of a distributing main for a general supply can requisition from the area board a supply to the premises (Electric Lighting (Clauses) Act 1899, Sched, s 27) unless a supply of energy is already given directly from electric lines belonging to the board or by a private supplier (Energy Act 1983, s 13). A 'squatter' in premises is not entitled to demand a supply under this provision (*Woodcock* v *South Western Electricity Board* [1975] 1 WLR 152). The statutory liability is to supply electricity. A board is not liable if loss or damage is caused because the supply is irregular or fluctuating, unless it makes a collateral contract with the consumer that the supply will at all times be adequate for a stated purpose (such as to keep infra-red lamps at a certain temperature so that chicks will not suffer: *Willmore* v *South Eastern Electricity Board* [1957] 2 Lloyd's Rep 375; in this case the claim failed, there being no collateral contract, although the board knew that chicks were being kept and made many attempts to regularise the supply). The board must lay any lines necessary to supply the maximum power but the cost of any line on the owner's or occupier's land and the cost of any line more than sixty feet from the main can be claimed by the board from the owner or occupier (1899 Act, Sched, s 27). A local authority may requisition a supply to any public lamps within seventy-five yards of a distributing main (Sched, s 29). A precedent for an agreement for such a supply is given in the *Encyclopaedia*, 4th ed, 23, 237.

3 Electricity Supply Regulations

The Electricity Supply Regulations 1937 (re-printed in 1972 with some amendments) have the object of ensuring the safety of the public and a proper and adequate supply of electrical energy. They deal with such matters as the constancy of the supply, the uniformity of pressure and the frequency of alternating current (regs 34, 35). A person may not use the current supplied to him in such a manner as to interfere unduly or improperly with the supply to any other consumer (Electric Lighting (Clauses) Act 1899, Sched, s 27(4)).

By s 16 of the Energy Act 1983 the Secretary of State may make such regulations as he thinks fit for the purpose of (*a*) securing that supplies of electricity by Electricity Boards and other persons are regular and efficient and (*b*) eliminating or reducing the risk of personal injury, or damage to property or interference with its use, arising from the supply of electricity by the Board or any other person, from the use of electricity so supplied or from the installation, maintenance or use of any electrical plant. By s 16(2) these regulations may relieve the Board from the duty to supply if electrical plant or fittings or their use do not comply with regulations; prohibit the supply of electricity except by a system approved by the Board; make provision for the keeping of maps, plans and sections by persons who supply electricity; and provide for other matters. No regulations had been made under s 16 up to 31 December 1985.

4 Powers of CEGB and area boards

4.1 Acquisition of land

The CEGB and area boards may acquire by agreement (Electricity Act 1947, s 2(5)), or compulsorily with the consent of the Secretary of State, any land required to carry out their functions (1947 Act, s 9). Compensation is governed by the provisions of the Land Compensation Act 1961, and the procedure is that laid down in the Acquisition of Land Act 1981, the Compulsory Purchase Act 1965, and the Compulsory Purchase of Land Regulations 1982 (SI No 6). The Secretary of State may authorise a board to purchase an easement or to purchase the right to place a line above or below ground without acquiring any other interest in the land (1947 Act, s 9) (this could be termed a statutory easement).

4.2 Lines over land

A board may not place an electric line above ground (except over its own land) without the consent of the Secretary of State and the local authority (1899 Act, Sched, s 10, as amended by the Electricity Act 1957, ss 31 and 32). The Electricity (Overhead Lines) Regulations 1970 (SI No 1355) apply. The board may obtain the agreement or views of the landowners over whose land the line is to run even before applying to the Secretary of State; this enables him to assess the extent of any likely objections. If the local planning authority objects there must be a public enquiry (1957 Act, s 34). At the same time as he considers the application to place an electric line the Secretary of State may consider an application for the compulsory grant of wayleaves to the board by private owners (Electricity (Supply) Act 1919, s 22). The Electricity Supply Regulations 1986 (SI No 1627) apply. If the board makes an application for these it must serve notice of its intention to apply on owners and occupiers concerned and they may make representations to the Secretary of State (ibid). He may grant the consent unconditionally or subject to such terms and conditions as he thinks fit; in considering the application he is to have regard to the effect on the amenities or value of the land of granting the application in the form proposed (ibid). If a local planning board objects to a proposal a public enquiry will be held and private objectors can be heard at the same time; private objectors cannot insist on a public enquiry, apparently, but the Secretary of State can order one if he thinks it advisable (1947 Act, s 66). The procedure to be followed at hearings on applications under s 22 is regulated by the Electricity Generating Stations and Overhead Lines (Enquiries Procedure) Rules 1987 (SI No 2182). Under the Town & Country Planning (Assessment of Environmental Effects) Regulations 1988 (SI No 1199) made under the EEC's Directive on Environmental Assessment (EEC/85/337) an environmental assessment of the possible impact on the environment of the granting of permission for any new power station of more than 300 MW thermal capacity will be needed. The Secretary of State may also call for such an assessment in relation to power stations of less than this capacity and for combined Heat and Power (CHP) or hydro schemes and overhead lines proposed by the CEGB or Area boards. The planning authority's views will be taken fully into account; see also DOE circular 34/76 (Welsh Office 45/76) on 'Procedure for consultation and consent for new electricity generating stations and overhead lines'.

When the course of the overhead line has been approved the board will enter into wayleave agreements with the owner or occupier. Copies of wayleave agreements can be obtained from the area offices of the boards, the CEGB and the Country Landowners' Association, 16 Belgrave Square, London, SW1. The *Encyclopaedia*, 4th ed Vol. 23, 276, contains a precedent for the grant of and consent to a wayleave.

The CEGB and area boards will supply particulars of agreements relating to the placing of cables above or below ground and of consents given but not yet acted on. This will be useful to solicitors, in rural areas particularly, who might otherwise overlook revealing these burdens to a proposing purchaser ((1968) *New Law Journal,* 18, 859).

The Electricity (Overhead Lines) Regulations 1970 (SI No 1355) govern such matters as the height, method of construction, materials and strength of overhead lines and their supports.

Express planning permission is not needed for the laying underground of pipes, cables and other apparatus, and the installation of service lines to individual consumers from an electric line, as these matters fall within the general permission given by the Town and Country Planning General Development Order 1988 (SI No 1813, in force from 5 December 1988) Sched 2, Part 17, Class G. By the conditions attached by Class G2 notice of certain operations must be given to the responsible specified authority before operations commence.

Compensation. Basic scales of compensation for interference with the farming and cultivation of land, and of rental (which can be annual or in a lump sum) for poles and towers, have been agreed by the National Farmers' Union and the Country Landowners' Association with the CEGB and area boards. The current agreed scale of payments for electricity wayleaves, negotiated between the Country Landowners' Association, the National Farmers' Union and the Electricity Supply Industry, runs from 1 August 1983 and may be found in CLA publication '*Electricity Wayleaves*' A 15/1987. Larger sums are paid for interference with commercial fruit growing than for interference with pasture. British Telecom (and presumably other licensed operators) has agreed to pay the same rates as the electricity industry. The entitlement to compensation for damage caused by the works arises under the Electric Lighting Act 1882, s 17. If there is a possibility that future development of the land may be curtailed by the lines, such as by reducing the number of houses which can be built on it when planning permission for residential development is obtained, provision should be made in the wayleave agreement for compensation in the future when this loss becomes apparent. In *Christian Salvesen Ltd* v *CEGB* (1984) 48 P & CR 465 it was held on the facts that a payment of £900 in 1966 for a wayleave easement for an overhead line and five pylons precluded the payment of later compensation when successors in title to the land were refused planning permission for residential development on the land by reason of the existence of the overhead line. The rights of the parties concerning compensation were governed exclusively by the deed. A similar case was *Mayclose* v *CEGB* [1987] 2 EGLR 18 where a deed of grant made by landowners with the CEGB provided for rent for wayleaves and also compensation for any diminution in the value of any land for which planning permission had been granted, by reason of the existence of overhead lines. A possible sterilisation of part of the land was envisaged but what happened was that a complete refusal of planning permission was issued. It was held that the owners could not have any more compensation, the deed was made after lengthy negotiations and was intended to encompass the whole of the relationship between the parties, including depreciation in value of the subject land. See also *Allen* v *South Eastern Electricity Board* [1988] 1 EGLR 171 (CA) where it was held on the facts that the phrase in a deed of compensation 'cannot take full advantage of the said planning permission' did not mean 'cannot sell the property for as high a price as could have been obtained but for the rights granted to the board'.

In view of these cases it would be wise for an owner to reserve carefully and unequivocally his right to future compensation, see a standard deed negotiated by the CLA and the NFU concerning gas pipelines. The boards will also pay for any injurious affection where trees have to be felled to make way for the wires, where land has to be left uncultivated because of them, or where minerals have to be left unworked because of the pylons. If lines prevent or impede the aerial spraying of crops, with increased expense for ground treatment, a claim can be made for this additional expense: *Pryor* v *Central Electricity Generating Board* (1968) 206 EG 1143 (£700 awarded on this account).

In *Clouds Estate Trustees* v *Southern Electricity Board* (Ref 183/1982) (1983) 268 EG 367 and 451 the annual rent paid under wayleave consents for overhead lines was assessed from 1 January 1982 at 50p per single pole, 50p per stay or strut, 75p per H pole and 25p per earthwire. The rent for underground cables was to be 15p per 25m run of cable and multiples of 25p and 30p per 50m run. Because of the permanency of the structures the assessment was effectively one by consent for the compulsory acquisition of rights over the land, despite the use of the term 'rent'. The land commanded the highest agricultural rents of any county in England, being in Wiltshire, and the award should not be taken as fixing a norm. The figures in use in 1987 included: 51p annually for a pole; 17p per 25 metres of cable, with much higher figures for interference with agricultural operations (arable £6.48 for a pole, £7.88 for a stay or strut).

4.3 Street works

Electricity boards may open and break up any street or bridge within their area and any sewers, drains and tunnels in such streets or bridges to lay down electric lines, conduits, service lines and other works, and to repair, alter or remove them (Electric Lighting Act 1882, s 12, incorporating earlier Acts, commencing with the Gasworks Clauses Act 1847). In this context 'tunnels' is construed ejusdem generis with sewers and drains, so a board is not entitled to open and break up a subway (*Schweder* v *Worthing Gas Light and Coke Co* [1912] 1 Ch 83). It may not break up any private street, railway or tramway without the owners' permission or the consent of the Secretary of State given after notice of the proposal and an opportunity to make representations has been given to the owners and occupiers (1882 Act, s 13). But a board may at any time enter upon any land where an electric line has been lawfully laid in order to replace, repair or alter it, and the consent of the owner, occupier or Secretary of State is not needed for this (1882 Act, s 12; 1847 Act, s 7). The board also has power to alter the position of any pipes or wires under any street or place which it is authorised to break up if these may interfere with the exercise of its powers, and other statutory bodies have a similar right in respect of electricity lines, subject in each case to following the procedure of due notice and payment of compensation (1882 Act, ss 15, 17). The body whose apparatus is to be affected may choose to carry out the proposed alterations to the apparatus itself at the proposer's expense (Electric Lighting (Clauses) Act 1899, Sched, s 17(*e*)).

The provisions of the Public Utilities Street Works Act 1950 (outlined in Chapter 10/4), govern the procedure to be followed where the proposed major works affect the local authority or other statutory bodies or persons. Plans and sections must be prepared and submitted to those to be affected (1950 Act, s 3). They may approve or disapprove, and disputes go to arbitration (ibid, s 3(6)). The board must reinstate streets or controlled land which it has broken up (ibid, ss 8, 18), and is liable for subsidence occurring within six months of reinstatement (ibid, s 7(6)). Liability continues beyond this period, however, for reinstatement negligently carried out (*Hartley* v *Rochdale Corporation* [1908] 2 KB 594). The board is entitled to receive payment for any alteration to or protection of apparatus made necessary by road or bridge alterations or transport works (ibid, s 22(1)).

Where a statutory undertaker causes a nuisance which is attributable to its carrying out of street works under statutory power, the authority is not liable, unless it has also been negligent (*Department of Transport* v *North West Water Authority* [1983] 3 WLR 707 (HL), construing Public Utilities Street Works Act 1950, s 18(2)).

Under the Electric Lighting (Clauses) Act 1899, Sched, s 18, where a board wishes to dig a trench for placing any new electric lines (except service lines) or other works near to any sewer, drain, watercourse or other works of the local authority or any main, pipe or siphon of any gas or water authority, it must (except in an emergency) give three days' notice to the appropriate one of these bodies and they may superintend the work and make reasonable requirements to protect their apparatus from injury; the board is liable for any damage caused. Similar provisions apply where the local authority or gas or water authority wishes to carry out works near lines or works of the electricity board (ibid). The electricity board must effectively insulate any of its electric lines which cross or are liable to touch mains, pipes, lines or services of the other authorities and must not lay lines so as actually to touch these mains, pipes, lines or services without the consent of the other authorities and the Secretary of State. If damage is caused by failure to comply with these requirements the board is liable to compensate the injured authority for it (ibid). Where an electric line has been placed above ground and any person lays a complaint before a magistrates' court that the line is or is likely to become dangerous to public safety, the court may order its removal upon such terms as it thinks fit (Electric Lighting Act 1882, s 14).

If damage is caused to electricity apparatus laid in a street by traffic of excessive weight, the board may claim compensation for the loss from the person responsible for the traffic (*Chichester Corporation* v *Foster* [1906] 1 KB 167: where a traction engine weighing upwards of 10 tons and drawing three trucks fractured a water main which had been sufficiently strong to withstand reasonable traffic for more than 30 years, the owner of the traction engine was held liable for the damage).

An electricity board must take all reasonable precautions

necessary to avoid injuring or affecting the working of telephone and telegraph wires (1899 Act, Sched, s 20). It must give seven days' notice of its intention to carry out works which may affect such lines, and the authorities affected may require the board to take protective precautions (ibid).

Chapter 13

Electricity After Privatisation

1 General background

The Electricity Act 1989 has the effect of privatising the electricity industry in England, Wales and Scotland (but except for certain sections it does not apply to Northern Ireland, s 103(2) and (3)). The Central Electricity Generating Board is split into two new generating companies (Powergen and National Power) and is required to make a scheme for the division of all its property, rights and liabilities between the two generating companies and a transmission company (the National Grid Company, which is to have the monopoly of high voltage electricity transmission). The Electricity Council is also to be required by the Secretary of State to make a scheme for the division of its property to a company or companies nominated by the Secretary of State (s 60(2)). Each of the 12 Area Boards will have its property transferred to new supply companies to be nominated by the Secretary (s 59) and in the case of the Scottish Boards their property will be transferred to three or more nominated companies of which two shall be designated as the Scottish electricity companies and one as the Scottish nuclear company. The system of regulation of generation, transmission and supply is in the hands of the Director General of Electrical Supply who has the functions of carrying out his duties in the way best calculated to promote competition in the supply and generation of electricity, to secure that all reasonable demands for electricity are satisfied, and to protect consumers' interests (s 3); some of these duties are to be exercised either by the Director or the Secretary of State (s 3(2) and (3). The Director is to establish consumers' committees for each area (s 2 and Sched 2); their functions are in ss 47 and 48.

The Director, in consultation with the Secretary of State and vice versa, deals with the issue of licences (discussed below).

2 Generation of electricity

It is an offence to generate electricity for the purpose of giving a supply of electricity to any premises or enabling a supply to be given, or to transmit electricity for that purpose, or to supply it to any premises, without a licence or an exemption from licensing (s 4(1)). 'Supply' in relation to electricity means supply through electric lines otherwise than to premises occupied by a licence holder for the purpose of carrying on his authorised activities; and 'transmit' means transmit by means of a system of high voltage lines and electrical plant used for conveying electricity from a generating station to a substation (s 4(4). Exemption may be granted from the need to have a licence for generation or supply but not from the need to have one for transmission (s 5(1) and (2)). Licences may include such conditions as appear to the grantor to be requisite and in particular may require the licence holder to enter into agreements for the use of his lines and plant by private electricity suppliers (s 7). Licences may be modified by agreement (s 10) and the Director may make reference to the Monopolies Commission for consideration whether any matters relating to the generation, transmission or supply under a licence operate or may be expected to operate against the public interest (s 11).

By s 33(1) a generating station shall not be constructed, extended or operated except in accordance with a consent granted by the Secretary of State except where the permitted capacity does not exceed 50 megawatts or such other capacity as the Secretary stipulates (s 33(2)). Consent is also required for the installation of overhead lines unless within the premises of the installer or of a nominal voltage not exceeding 20 kilovolts (s 34(1) and (2)).

3 Transmission

It shall be the duty of the holder of a licence authorising him to transmit electricity to develop and maintain an efficient, co-ordinated and economical system of transmission and to facilitate competition in the supply and generation of electricity (s 8(2)). The holder of a transmission licence will be under conditions as to price, connection charges and use of the transmission system on a non-discriminatory basis and will be required to make available price and technical information to enable prospective users to assess the options and likely costs of connecting to the grid (ss 41–44 and s 89).

4 Supply

A public electricity supplier (PES) is a person who is authorised by a licence under s 6(1)(*c*) to supply electricity to any premises in his authorised area (s 6(9)). It is the duty of a PES to develop and maintain an efficient, co-ordinated and economical system of electricity supply (s 8(1)). A PES shall, subject to the Act and regulations, upon being required to do so by the owner or occupier of any premises, give a supply of electricity to those premises and so far as may be necessary for that purpose, provide electric lines or electric plant or both (s 15(1)). A person requiring a supply shall give due notice and as soon as practicable thereafter the PES shall give him notice stating the extent to which the request is acceptable, the prices to be charged, the payment or security required and any restrictions which must be imposed under regulations made under s 27 (which relate to supply and safety) (s 15(2) and (4)). There is no duty to supply if there is a supply being given to the premises by a private electricity supplier (s 16(1)). The Director will determine any dispute (s 21(1)).

The public electricity supply code is contained in Schedule 6 and deals with: the recovery of electricity charges; the restoration of supply after it has been cut off as a result of a default by a tariff customer; the offence of restoring the supply without consent; the intentional or culpably negligent damage of electrical plant; powers of entry by authorised officers of the PES for inspection, discontinuance, replacing or repair; duties of officers to safeguard the premises after authorised entry.

By s 18(1) where any electric line or electrical plant is provided by a PES in pursuance of s 15(1) the supplier may require any expenses reasonably incurred to be defrayed by the person requiring the supply, to such extent as is reasonable in the circumstances. The reference to 'expenses reasonably incurred' includes also a reference to the capitalised value of any expenses likely to be incurred by the supplier in maintaining the line or plant in the future in so far as they will not be recoverable by the supplier as part of the charges made by him for the supply (s 18(4)). The Secretary of State may, after consultation with the Director, make regulations for entitling the supplier to claim expenses reasonably incurred in providing a line or plant (s 18(2)). A person who requires a supply of electricity may enter into an agreement for it where the maximum power to be made available exceeds 10 megawatts or such other wattage as the Secretary of State may by order specify and it is otherwise reasonable in all the circumstances for such an agreement to be entered into; the Secretary of State must first consult with public electricity suppliers and others appearing to him to be representative of persons likely to be affected (s 20(1) and (2)). The Director determines disputes under ss 15 to 20 between a PES and a person requiring a supply of electricity (s 21(1)).

By Sched 7 clause 1, where a customer of an electricity supplier is to be charged wholly or partly for a supply by reference to the quantity of electricity supplied the supply shall be given through a meter supplied by the electricity supplier (or if agreed, by the customer). Meters are to be certified by meter examiners appointed by the Director of Electrical Supply (para 4(1) and 5) and it shall be the duty of the meter examiner to examine and test any meter at the request of the customer (para 7). It is a criminal offence (summary conviction can bring a fine not exceeding level 3) if any person intentionally or by culpable negligence alters the register which records the quantity of electricity supplied or prevents the meter from registering the quantity (para 11(1)).

5 Powers of licence holders

Section 9 provides that Scheds 3 and 4 (which relate to compulsory acquisition and other powers) shall have effect in relation to a PES or a person authorised by licence to transmit electricity and, to the extent that the licence so provides, in relation to any other licence holder.

5.1 Acquisition

By Sched 3, para 1(1) the Secretary of State may authorise a licence holder to purchase compulsorily any land required for any purpose connected with the carrying on of the activities authorised by the licence. By para 1(2) 'land' includes 'any right over land (other than, in Scotland, a right to abstract, divert and use water); and the power of the Secretary of State under this paragraph includes power to authorise the acquisition of rights over land by creating new rights as well as acquiring existing ones'. A consent to acquire which is not acted on within six months of the day that it is granted ceases to have effect (para 1(3)). Where a licence holder has acquired land under para 1 he shall not dispose of it or of any interest in it without the consent of the Director (para 1(4)). By Part II of Sched 3 the procedure on acquisition is that applicable under the Acquisition of Land Act 1981 (para 5(1)) and paras 6 to 14 give appropriate adaptations of the Compulsory Purchase Act 1965 to enable new rights to be acquired.

5.2 Street works

Schedule 4 confers other powers on licence holders. By para 1 a licence holder may: install works under, over, in, on, along, or across any street and from time to time inspect, maintain, adjust, repair, alter, replace or remove any electric lines or plant and any structures for housing or covering any such lines or plant. For those purposes the licence holder may open and break up any street or any sewers, drains or tunnels within or under any street, or tunnel or bore under any street or remove or use all earth and materials under any street (para 1(1)). By a proviso to para 1 he may not 'place any electric line or electric plant into, through or against any building, or in any land not dedicated to public use'. The power to place on or over a street shall be exercisable only with the consent of the highway authority (in the case of a publicly maintainable street), or with the consent of the railway or navigation authority (where the street is under the control of one of them), or, if the street does not fall under one of them but is repairable by a person, with the consent of that person. In all three cases the consent is not to be unreasonably withheld (para 1(2)). A licence holder shall do as little damage as possible in exercising his powers under para 1 and shall make compensation for any damage done in exercise of them (para 1(6)). He shall so exercise his powers as to secure that nothing which he installs or keeps installed under, over, in, on, along or across any street becomes a source of danger to the public (para 1(7)).

A licence holder may execute works under paras 1 or 2 notwithstanding that they involve a temporary or permanent alteration to: any electric line or electric plant under the control of another licence holder; any gas pipe under the control of a public gas supplier; any water pipe under the control of a person supplying water under the exercise of statutory powers; or any telecommunication apparatus operated by a person to whom the telecommunications code applies. The licence holder must give to the 'relevant undertaker' not less than one month's notice of the proposed work and the time and place of it (para 3(3)), except in case of an emergency (para 3(4)), and the relevant undertaker may within seven days give a counter-notice saying that he intends to do the necessary works or that they must be done under his supervision; he can charge the licence holder with the expenses if he does the work himself (paras 3(5) and (6)). Any public gas supplier, water supplier under statutory powers or public telecommunications operator who is authorised to execute works corresponding to those authorised by paras 1 or 2 may carry out his authorised works notwithstanding that they involve a temporary or permanent alteration to any electric line or electrical plant under the control of a licence holder (para 4(1)).

5.3 Wayleaves

By para 5 if it is necessary or expedient for a licence holder in connection with carrying out his authorised activities to install and keep installed an electric line on, under or over any land he may serve notice to that effect on the owner or occupier of the land. If the owner or occupier has failed within the period specified in the notice (not being less than 21 days) to give consent, or has given it on terms to which the licence holder objects the licence holder may apply to the Secretary of State who can grant the wayleave on such terms and conditions as he thinks fit, after affording the owner or occupier an opportunity to be heard (paras 5(4) and (5)). A necessary wayleave granted under para 5 shall bind any person who is at any time the owner or occupier of the land (para 5(6)(b)). If the wayleave is determined (for example by the effluxion of the time for which it was granted) the owner or occupier of the land may give notice to remove the electric line but the licence holder need not comply if he either applies for a wayleave under para 5 or applies for an order for compulsory acquisition of the land (para 6(3)); otherwise he must remove the line within 3 months from the date of the notice (para 6(3)). There is an appeal to the Secretary of State (para 6(4) and (5)).

5.4 Felling and lopping trees

Paragraph 7(1) of Sched 4 applies where any tree is or will be in such close proximity to an electric line or electrical plant which is kept installed or is being or is to be installed by a licence holder as to obstruct or interfere with its installation, maintenance or working or as to constitute an unacceptable source of danger (whether to children or to any other persons). In such a case the licence holder may give notice to the occupier of the land requiring him to fell or lop the tree or cut back its roots so as to prevent it from having the effect mentioned above, subject to the licence holder paying the expenses reasonably incurred in so doing (para 7(1) and (2)). If within 21 days of the giving of such a notice the owner or occupier gives a counter-notice objecting to the requirements of the notice the matter shall be referred to the Secretary of State if the counter-notice is not withdrawn (para 7(5)). On such reference the Secretary of State may empower felling or lopping or cutting back of roots to be done and may determine the question of the expenses (para 7(6)). Where the licence holder exercises powers under para 7(4) or (6) he must cause trees to be felled or lopped or their roots to be cut back in accordance with good arboricultural practice so as to do as little damage as possible to the tree and other trees, hedges, fences and crops; he must cause the cuttings to be removed as the owner or occupier may direct and he must make good any damage to the land (para 7(7)). In para 7 'tree' includes any shrub (para 7(8)). It is an offence for any person intentionally to obstruct a person acting in the exercise of any power conferred by paras 7 and 8, (para 9(1)).

6 Repeals

There are many repeals of words or sections of various Acts, in many cases the word 'electricity' is to be deleted from the definition of 'statutory undertaker'. The following Acts are entirely repealed by s 102(4) and Sched 16 (the list is not exhaustive): The Electric Lighting Act 1882, The Electric Lighting (Clauses) Act 1899, The Electric Lighting Act 1909, The Electricity (Supply) Act 1919, The Electricity (Supply) Act 1926, The Electricity Supply (Meters) Act 1936, The Electricity Act 1947, The Electricity Act 1957, The Electricity and Gas Act 1963, The Gas and Electricity Act 1968, The Electricity Act 1972, The Electricity (Scotland) Act 1979, The Electricity (Financial Provisions) (Scotland) Act 1988.

Chapter 14

Gas

The Gas Act 1986 'privatised' the British Gas Corporation. It provided for the issue of shares in a new company, British Gas plc, and for the vesting in this company of the property, rights and liabilities of the British Gas Corporation from 24 August 1986 (s 49). The BGC was dissolved by s 57. The former monopoly position of BGC as a supplier of gas was abolished (s 3) and gas may now be supplied by any 'public gas supplier' (PGS) who holds an authorisation under s 7, or anyone authorised by the Secretary of State, s 8. It is an offence to supply gas through pipes to any premises unless the supplier is authorised under ss 7 or 8 or comes within the exceptions set out in s 6 (s 5). A Director General of Gas Supply ('the Director') is appointed by the Secretary of State to perform the functions assigned to him by ss 4, 34–38 and Sched 1 (s 1). His functions include protecting the interests of consumers and protecting the public from dangers arising from the transmission of gas through pipes (s 4). A Gas Consumers' Council replaces the national and regional Gas Consumers' Councils (ss 2, 4, 41 and Sched 2).

Under the 1986 Act it is the duty of a PGS to develop and maintain an efficient, co-ordinated and economical system of gas supply, (s 9), and to avoid any undue preference in the supply of gas to persons entitled to a supply under s 10 (below). Subsequent references in this chapter are to the 1986 Act unless otherwise stated.

1 Duty to supply

An owner or occupier may require from a PGS a supply of gas to any premises which are within 25 yards of a relevant main of the supplier or are connected by a service pipe to such main (s 10(1)) and in the former case the PGS shall also provide and lay any necessary pipe but the PGS is not under an obligation to supply to any premises gas in excess of 25,000 therms in any period of twelve months (s 10(5)). No PGS can be required to give or continue a supply if prevented by circumstances beyond its control, or if there could be danger to the public by so doing (s 10(6)). If a PGS agrees to supply more than 25,000 therms yearly to any premises it may do so by special contract (s 10(7)).

The cost of providing and laying any pipe falls upon the person requiring the supply under s 10(1) to the extent that it runs over his land or, to the extent that it does not run over his land, that it exceeds a distance of thirty feet from the main (s 10(3)). These dimensions in yards and feet have survived the Gas (Metrication) Regulations 1980 (SI No 1851) made under an EEC Directive.

The PGS is under a duty to maintain, repair or renew the pipe at the reasonable expense of the owner or occupier (Sched 5, para 1). A person requiring a supply must give reasonable notice to the PGS and if so required give security for payment of expenses, s 10(2) and 11. The PGS's obligation to supply does not extend beyond a supply for domestic purposes or lighting if the capacity of the main is insufficient to supply gas for any other purpose, s 10(7). But if a person requiring a supply for some other purpose, such as an industrial one, contracts with the PGS so as to justify the PGS laying down the mains or other apparatus for the larger supply, then the person will become entitled to a supply, s 10(7).

Where the supply of gas to premises is cut off or is likely to be cut off for failure to pay for the supply the local authority may, at the request in writing of the occupier, make an arrangement with the PGS for restoration or continuance of the supply on agreed terms, Local Government (Miscellaneous Provisions) Act 1976, (as amended) s 33.

Section 15 and Sched 5 contain the Public Gas Supply Code. In addition to the matters already mentioned it deals with meters, (paras 3–6), improper use of gas and injury to gas fittings (paras 9–11), the duty of a PGS to prevent escapes of gas (para 13), entry onto premises to inspect, discontinue, supply, replace or alter pipes (paras 15–18).

2 Right to lay pipes

In order to supply gas to a person whom they are bound or entitled to supply (or in order to receive a bulk supply), the

Corporation may break up and open any street or bridge (Sched 4, para 1). The Corporation must do as little damage as possible and make compensation for all damage done (ibid). Loss of profit to the business of a person affected by the works is possibly included in the term 'damage', despite *Argyle Motors Ltd* v *Birkenhead Corporation* [1975] AC 99. See *Leonidis* v *Thames Water Authority* (1979) 77 LGR 722. The Corporation may also open and break up any sewers, drains or tunnels in or under any street or bridge and lay pipes, conduits and service pipes, and repair, alter or remove them (ibid) but it cannot erect a structure for housing pressure governors in a street or bridge without the consent (not to be unreasonably withheld) of the highway authority (ibid). A board cannot lay its pipes in private land without the consent of the owner and occupier (ibid), but it may enter on private land, after giving not less than seven days' notice in writing to the owner and occupier, to repair, replace or alter any pipe lawfully laid there (ibid). No prior notice is required in a case of emergency but notice must be given as soon as possible after the emergency (ibid).

The laying underground of mains, pipes or other apparatus by the Corporation does not need an express planning permission as it falls within the Town and Country Planning General Development Order 1988 (SI No 1813, in force from 5 December 1988) Sched 2, Part 17, Class F. By the conditions attached by Class F2 prior notice must be given to the responsible planning authority before operations commence.

An authorised officer of a PGS may enter on premises at all reasonable times, on showing his documents of authority, to inspect meters, fittings and works for the supply of gas (Sched 5, para 15). Premises may be entered after 24 hours' notice to the occupier (or to the owner or lessee if the premises are unoccupied) for the removal of meters, pipes and fittings, where the occupier ceases to require a supply or where a PGS is authorised to cut off the supply (ibid, para 16). See the Gas Safety (Rights of Entry) Regulations 1983 (SI No 1575).

3 Compulsory purchase

A PGS may be authorised by the Secretary of State to purchase compulsorily any land (including any right over land, and to create and acquire new rights) which is required for or in connection with the exercise and performance of its functions (s 9(3), Sched

3, para 1). The procedure laid down in the Acquisition of Land Act 1981, Sched 1, and the Compulsory Purchase Act 1965, applies (Sched 2, paras 4 and 5). Notices must be served on the owners and occupiers and advertisements made. They can object and the Secretary of State may then hold a public enquiry. Compensation is payable, assessed by the Lands Tribunal if agreement cannot otherwise be reached, at an open market value under the Land Compensation Act 1961, s 5. Compensation is also payable for injurious affection, severance or disturbance, just as in the case of the laying of sewers, already discussed. In *Wells-Kendrew* v *British Gas Corporation* (1974) 229 EG 272, the following sums were paid: 12½p per linear yard for the laying of an 18-inch natural gas pipe for 211 yards in woodland in a 20 foot wide strip; 25p per linear yard for the pipe-line for a 20 foot wide strip 541 yards long in pasture land. The total payment was £162.50. In *Vaynor Developments* v *Wales Gas Board* [1977] CLY 332 the claimants bought land with planning permission for fifteen houses, subject to the Water Board's rights over a 40 foot wide strip. Shortly afterwards the Board extended rights to a 320 foot strip. Because of this the claimants abandoned negotiations with builders for a development of sixty-nine houses for which planning permission was virtually assured. The claimants asked for £51,000 on the basis that they had nearly reached agreement for the development; the Board offered £11,500 on a sale-price basis. The sum of £27,000 was awarded on a 'ripe for development' basis but the higher claim was discounted for uncertainty, risk and deferment. Compensation may need to be negotiated for such things as the disturbance and checking of trees when a main is laid under or near to them, and for minerals which must remain in the ground unworked because of the danger that working would cause subsidence to the mains.

Under a national agreement between the Country Landowners' Association, the National Farmers' Union and the Gas Industry, rates of compensation for such things as the grant of an easement for a gas main, disturbance payments, loss of development value, and the mining of minerals, have been agreed. The rate will vary according to the value per acre of the land affected and the width of the strip affected. Lump sum, once-for-all payments are made, ranging from £2.48 per yard run for an 80 foot strip over land worth £600 per acre, to £12.40 per yard for the same strip over land worth £3,000 per acre (CLA Subscribers' Circular 2/80, dated 15 April 1980, contains details). It is understood from the Country

Landowners' Association that for gas pipelines the CLA and the NFU try to negotiate at the start of a gaslines project for a capital sum relating to the whole length of the pipeline, though individual owners may negotiate for themselves if they wish over such land as, for example, market gardens, organic farms and development land. The standard deed of grant negotiated by the CLA and the NFU makes the pipeline operator responsible for any drainage problems which arise as a result of the pipeline, in perpetuity, and also permits further compensation to be paid as and when development is frustrated because of the pipe.

4 Street works

The Corporation is subject to the Public Utilities Street Works Act 1950 (outlined in Chapter 10/4). If works proposed by the Corporation in the street or in controlled land will affect the apparatus of other authorities, they must be supplied with a plan and sections (1950 Act, s 3). They are entitled to object and to have any dispute arbitrated (ibid, s 4). The Corporation must reinstate the street and is liable for any subsidence occurring within six months of the works (ibid, s 7). Under Pt II, ss 21–23, of the 1950 Act the Corporation is entitled to payment for any works rendered necessary to its apparatus by road, bridge or transport works of other authorities.

5 Nuisance by a PGS

A PGS shall do as little damage as possible in the exercise of its powers and shall make compensation for damage done in such exercise (Sched 4, para 3). It is not liable for nuisance if it did not cause or adopt the nuisance (*Dunne* v *North Western Gas Board* [1964] 2 QB 806 (CA)). A PGS is liable criminally if it fails within 12 hours of receiving notice of the escape of gas to prevent it unless it can show that compliance was not reasonably practicable within the time and that it did comply as soon as it was reasonably practicable to do so (Sched 5, para 13).

Where a statutory undertaker causes a nuisance which is attributable to its carrying out of street works under statutory power, the authority is not liable unless it has also been negligent (*Department of Transport* v *North West Water Authority* [1983] 3 WLR 707 (HL), construing the Public Utilities Street Works Act 1950, s 18(2)).

6 Damage to property of a PGS

A person who intentionally or by culpable negligence injures or allows to be injured any pipes, meter or fittings of a PGS may be prosecuted (Sched 5, para 10).

The Corporation may decline to supply gas to any person who improperly uses it or so deals with it as to interfere with the efficient supply to any other consumer (Sched 5, para 9).

A person will be liable to the Corporation for damaging pipes properly laid in streets by reason of extraordinary traffic such as a very heavy load (*Gas Light & Coke Co* v *St Mary Abbott's, Kensington, Vestry* (1885) 15 QBD 1: gas pipes under a road were damaged by a heavy steamroller carrying out street repairs—a claim for the damage was successful).

7 Installation of gas equipment

A person who installs or alters gas equipment in or on premises must do so in accordance with the safety standards laid down in the Gas Safety Regulations 1972 (SI No 1178), as amended by SI 1882, 1851, 1358. See also the Gas Safety (Rights of Entry) Regulations 1983 (SI No 1575).

8 Underground storage of gas

The Gas Act 1965 authorises a PGS to store gas in natural porous strata underground provided they have a storage authorisation order from the Secretary of State (1965 Act, s 4). He may at any time impose conditions on the manner of storage if he considers it necessary to do so in the interests of safety (ibid, s 16), and a PGS is absolutely liable in civil proceedings for damage caused by gas in underground gas storage or in boreholes connected with underground gas storage or by an escape of gas from storage or boreholes (ibid, s 14). Compensation is payable where percolating water is polluted by the underground gas, whether stored or escaped (ibid, s 15). In addition to designating an area for storage, the Secretary of State may designate an area (called 'the protective area') within which excavating, mining, quarrying or boring below a prescribed depth is controlled so as not to cause the danger of escapes of gas (ibid, s 5). Particulars of the storage area and protective area are registrable as local land charges (ibid, s 5(10)). A person who proves that his interest in land is depreciated in value by reason of the making of a storage authorisation order is entitled to compensation equal to the amount of the depreciation (ibid, s 7). The compensation is assessed by the Lands Tribunal (ibid, s 23) if the parties do not agree it.

The form of the storage authorisation order and the procedure for applying for it are set out in Sched 2 to the Gas Act 1965. By para 1 of Sched 2 a storage authorisation order shall specify by reference to a large-scale map attached to the order the extent and location of the stratum (and the surface perimeter) in which gas may be stored ('the storage area'); shall state the nature of the gas authorised to be stored; and shall specify the protective area.

When applying for an authorisation order the applicants must first supply written proposals giving such details as the perimeter of the storage area and protective area, the depth of proposed storage, the nature and volume of gas, the proposed method of injecting and withdrawing gas and particulars of exploratory work done and results obtained (Gas Act 1965, Sched 2, para 3(2)). Notice of the proposals must be published on two successive weeks in the Gazette and local newspapers circulating in the area, saying where the detailed proposals and map may be inspected (ibid, para 4(1)). Copies of the notice must be served on every local planning authority, local authority and water authority in the area and on such other body or person as the Secretary of State may direct (ibid, para 4(2)). He will then decide whether to allow a formal application or not; he may allow it to proceed with modifications (ibid, para 5). The formal application to the Secretary of State will give a specification of the surface works, boreholes and pipes; a large scale map showing the perimeters of the storage area and protective area; sufficient particulars of any surface works; a statement of depth of storage; and particulars of the nature and volume of gas to be stored (ibid, para 6(1)). There must be notice by advertisements again in the Gazette and local newspapers on two successive weeks; the notice will state (inter alia) that objections to the Secretary of State must be made not less than 28 days from the date of first publication of the notice (ibid, para 7(2)). A copy of the notice must be served on every local planning authority, local authority, river authority, statutory water undertaker and highway authority in the area affected, on every owner, lessee and occupier (except tenants for a month or less) within the storage area or protective area and on such other persons or bodies as the Secretary of State may direct (ibid, para 7(3)). If an objection is made which is not trivial or frivolous and does not merely relate to compensation, he must hold a public local enquiry (ibid, para 8). The procedure is governed by the Gas (Underground Storage) (Inquiries Procedure) Rules 1966 (SI No 1375). If the order is made, the applicants must publish in the Gazette and local newspapers a notice to that effect and must serve the notice on every person who was entitled to be notified of the application to the Secretary of State (ibid, para 10 (6), (7)). A copy of the order and map must be lodged with each local authority (ibid, para 10(7)).

A PGS may obtain by compulsory purchase the right to store gas in an underground gas storage (Gas Act 1965, Sched 4, Pt I). Liquified natural gas is stored in six major installations and in five salt cavities near Hornsea (British Gas Annual Report 1988, p 8). The Secretary of State may also authorise a PGS under para 1 of Sched 3 to the 1986 Act to purchase compulsorily any land which is in a storage area or protective area which is the site of any well, borehole or shaft, in order to use the well, borehole or shaft in connection with the underground gas storage or (where it goes below the depth at which the gas is stored) to block it up or stop anyone else from using it (1965 Act, s 13(1)). A right of way for access to land on which there is, or is to be made, a well, borehole or shaft may be compulsorily granted also, and the grant of this will carry with it any necessary ancillary rights, such as to construct and maintain gates, stiles and bridges along the way (ibid, s 13(3)).

Chapter 15

Street Works

1 Highway authorities

Under s 1(2) (as amended) of the Highways Act 1980 each county council or metropolitan district council is the highway authority outside Greater London for highways which are not trunk roads or other special roads. For trunk or special roads the highway authority is the Secretary of State for the Environment though he may enter into an agreement for the county council to exercise some of his functions (and they, with his consent, may agree for a district council to exercise). In London the highway authority is the metropolitan district council (it was formerly the Greater London Council) for metropolitan roads and the appropriate non-metropolitan council for highways (s 2(3)). Outside London the district council may maintain public highways (s 42(1)). The Greater London Council and the metropolitan county councils were abolished from 1 April 1986 by the Local Government Act 1985, s 1.

2 The making up of private streets

Where a street works authority considers that any private street within its area is not sewered, levelled, paved, metalled, flagged, channelled, made good and lighted to its satisfaction it may resolve to carry out street works and, subject to the private street works code, the expenses so incurred shall be apportioned between the premises fronting the street (Highways Act 1980, s 205(1)). By s 203(1) 'the private street works code' means ss 205–218.

By s 329(1) the term 'street' includes any highway, road, lane, footpath, square, court, alley, or passage whether a thoroughfare or not, and any part of a street. By s 328(1) 'highway' means the whole or part of a highway, other than a ferry or waterway, except where the context otherwise requires.

By s 206 a street works authority may include in street works to be executed any works which the authority thinks necessary for bringing the street, as regards sewerage, drainage, level, or other matters, into conformity with other streets, including the provision of separate sewers for the reception of sewage and of surface water respectively. Under the 1980 code the proper officer of the road authority prepares specifications, makes an estimate of probable cost and a provisional apportionment of cost 'between the premises liable to be charged with them' (s 205(3)), that is, those 'fronting the street' (s 205(1)). By s 203(3) 'fronting' includes 'adjoining'. By s 207(3) the authority may include in the apportionment premises which do not front the street but have access to it through a court, passage or otherwise if they will in the authority's view be benefited by the works; the contribution will be fixed according to the degree of benefit to be derived by the premises. The apportionment upon premises liable to be charged is in general made 'according to the frontage of the respective premises' (s 207(1)). A person can be a frontager even if his premises are separated from the street by, for example, a strip of waste land (*Warwickshire County Council* v *Adkins* (1967) 112 SJ 135). The owner of a first floor maisonette was held not to be a frontager in *Buckinghamshire County Council* v *Trigg* [1963] 1 WLR 155.

After the provisional apportionment has been made the authority resolves to adopt the specifications and public advertisement of the scheme is made (s 205(5)). Copies are served on the owners affected and they may object within one month of publication (s 208). Objections are heard by the magistrates' court (s 209), which may quash in whole or in part or may amend the resolution of approval, specification, plans, sections, estimate and provisional apportionment or any of them (s 209(2)). After the scheme has been approved and objections have been dealt with the work is carried out by or for the highway authority and the cost is recoverable from the owners for the time being of the premises (s 212(1)), after a final apportionment has been made under s 211 and notice has been served on the owners (ibid).

If the owners of property accounting for more than one-half of the frontage of a made up street (or part not less than 100 yards in length) request it, the highway authority must declare the street to be repairable at the public expense, provided that security has been given under s 219 and not refunded or released (s 229).

By s 37 a highway may become repairable at the public expense by dedication by any person, if the authority does not decline to accept it. By s 38 a highway authority may adopt a highway by agreement with a person and thereupon the liability of that person to maintain will be extinguished. By s 41(1) if a highway is maintainable at the public expense the highway authority is under a duty to maintain the highway.

By s 69(1) of the Housing Act 1988, when any street works have been executed in a private street in an area designated as the area of a housing action trust, the trust may serve a notice on the street works authority requiring it to declare the street a highway maintainable at the public expense. The highway authority may within two months of the service of the notice appeal to the Secretary of State on grounds relating to its construction, design, layout or maintenance (s 69(2)). The Secretary may determine the appeal and may impose conditions on the trust, on compliance with which the notice will take effect (s 69(3) and (4)).

2.1 Byelaws concerning new streets

The county council has power to make byelaws regulating the level, width and construction of new streets and the provision for the sewerage of such streets (Highways Act 1980, s 186(1)). In London the metropolitan district councils and the Common Council, as appropriate, exercise the power (ibid). Any plans for the works required to make up new streets must be rejected by the authority if they do not comply with the byelaws (s 191). The Secretary of State may order the authority to make byelaws (s 186(1)) and make them himself if they are not made by the authority within three months of his order (s 186(3)). New street byelaws may include provision as to (inter alia) the inspection of work, the testing of sewers and the taking by the county council or London metropolitan district council of samples of materials to be used in the execution of the works (s 186(5)).

2.2 Statutory undertakers

The Highways Act 1980 contains in s 156 provisions designed to prevent (so far as emergencies permit) highways from being taken up successively by statutory undertakers who have not proper liaison with each other. A road may not be taken up again within twelve months of its having been more than one-third in width obstructed by works or one-third or more in width resurfaced. An exception to this prohibition is where work is required to be done to provide a service pipe or overhead telecommunication apparatus to premises not already supplied (s 156(4)(*a*)), or in an emergency (s 156(3)).

2.3 Drainage onto the highway

A local authority or highway authority may serve notice on the occupier of premises adjoining a highway requiring him within twenty-eight days to construct such channels, gutters or downpipes as will prevent water falling from his roof or any other part of the premises onto people using the highway and as will prevent (so far as reasonably practicable) surface water from flowing onto or over the footway (Highways Act 1980, s 163(1)).

2.4 Bridging a highway

The highway authority may grant a person a licence to construct a bridge over a highway on such terms and conditions as it thinks fit. It cannot charge him for the licence, except for legal or other expenses, but he must remove the bridge when the authority requires him to do so, if the authority needs to improve the highway (Highways Act 1980, s 176).

2.5 Building over a highway

By s 177(1) of the Highways Act 1980 except in the exercise of statutory powers no building may be constructed over any part of a highway maintainable at the public expense, without a licence from the highway authority; and no building constructed or altered under such a licence shall be used otherwise than in accordance with the terms and conditions of the licence.

2.6 Pipes over highway

No person may without the consent of the highway authority (which it may grant on such reasonable terms and conditions as it thinks fit) fix or place any overhead beam, rail, pipe, cable, wire or other similar apparatus over, along or across a highway (Highways Act 1980, s 178(1)). Appeal on a refusal or as to the reasonableness of conditions is to a magistrates' court (s 178(2)).

2.7 Vaults and cellars under highway

No one without the consent of the appropriate authority may construct any part of a building or a vault, arch or cellar under a carriageway of a street (Highways Act 1980 s 179(1)), nor may anyone without consent make an opening in a footway as an entrance to a vault or cellar (s 180(1)). An authority may by notice require works to be removed, altered or dealt with as specified in the notice if they contravene s 179 (Local Government (Miscellaneous Provisions) Act 1982 s 22(1)). Every vault, arch and cellar under a street and every opening from the street into such works shall be kept in good condition and repair by the owner or occupier (s 180(6)). A breach of s 180(6) does not of itself give rise to a civil action; there must be negligence or nuisance by the person responsible for repair before an action will lie. In *Scott* v *Green & Sons* [1969] 1 WLR 301 (CA) a builder's lorry cracked a flagstone over the defendants' cellar, unknown to the defendants. Minutes later the plaintiff fell into the cellar and was injured. The defendants were held not liable. But where a person was injured because her leg went through a defective grating of which the occupier knew, he was liable in nuisance, *MacFarlane* v *Gwalter* [1958] 1 All ER 181.

3 The making of roads by developers

The system of payment for work to be done is known as 'the advance payments code' and by s 203(1) of the Highways Act 1980 the phrase means ss 219–225.

By s 219(1), where it is proposed to erect a building for which plans are required to be lodged with the local authority for building regulations permission and the building will front onto a private street for which the street works authority will have power under the relevant street works code to require works to be executed or to execute works, no work shall be done to erect the building unless a sufficient sum for the cost of the street works has been paid to the authority or has been secured to the satisfaction of the authority. Certain exceptions, such as where the street is not joined, or likely to be joined for a reasonable time, to the public highway, are given in s 219(4). The street works authority then calculates what sum would be recoverable if the authority itself carried out such work as the authority would require before declaring the street to be one maintainable at the public expense (s 220(3)). Within six weeks of passing the developer's plans the authority serves notice on him requiring the payment or the securing of the sum specified in the notice (s 220(1)). He can then deposit the sum with the authority or, much more usually, enter into an agreement with the authority under s 38. Under such an agreement the developer agrees to construct the roads within a stated time to the standard laid down by the authority and the authority in turn agrees to maintain the works when they have been completed and take them over in a stated time, for maintenance at the public expense. The s 38 agreement is almost invariably supported by a bond, in which an insurance company, bank or similar institution guarantees due performance by providing the money for completion of the works if the developer defaults.

The authority which enters into s 38 agreements differs with the area and the nature of the road. By s 1, for example, the Minister is the highway authority for trunk roads but by s 6 he may delegate his functions to a county council, a metropolitan district council or a London borough council; outside the former GLC area the county council is the highway authority for all except trunk roads but by s 42 the non-metropolitan district council may undertake maintenance of any eligible highway in the district and under this many s 38 agreements are made by the non-metropolitan district council.

The making up of the sewers for the new roads falls within the Public Health Act 1936 and the equivalent of a s 38 agreement is a s 18 agreement under the 1936 Act. This is mentioned more fully in Chapter 6/2.

Where land is in the area of an urban development authority established under the Local Government, Planning and Land Act 1980 special considerations apply to the grant of planning permission for making roads and the acquisition of land for the building of roads to service the area, see part XVI of the 1980 Act and particularly ss 136, 148 and 150.

4 Street works by public utilities

The Public Utilities Street Works Act 1950 lays down in Pt I a code of procedure for statutory undertakers to regulate work involving breaking up streets and operations on sewers, drains and tunnels under streets and bridges. The code relates to any statutory power to carry out 'undertakers' works' and these are defined as works for purposes other than road purposes of either of the two following types: firstly, placing, inspecting, maintaining, adjusting, repairing, altering, renewing, changing the position of or removing apparatus, and secondly breaking up or opening or tunnelling or boring under a street or controlled land for the above purposes, breaking up or opening a sewer, drain or tunnel for those purposes, and other works requisite for or incidental to those purposes (s 1(2)). Controlled land is, inter alia, land abutting on a street which is, or is prospectively, a maintainable highway (Sched 1, para 1(1)).

None of these undertakers' works may be carried out until a plan and section have been agreed between the undertakers and the street authority, or sewer authority or person having management or control of a private street, as the case may be (s 3). The Act calls them the street authority or street managers (s 2). Under s 3(2) no such formalities are required for such minor matters as inspecting, maintaining, adjusting or repairing apparatus (as defined in s 39(1)); placing, altering, renewing, or changing the position of a service pipe or service line (such as a gas pipe from the main to a house, or a private sewer or drain, or an underground telephone line serving individual premises) or an overhead telegraphic line; placing apparatus required in connection with building work which will be removed on completion of the work; and any breaking up, opening, tunnelling or boring required in connection with the above.

The street authorities or street managers to whom plans and sections are submitted may by notice to the undertaker approve, disapprove or approve subject to modifications and must state their grounds in a case other than one of approval (s 4). Any dispute goes to an arbitrator appointed by agreement or in default of agreement by the President of the Institution of Civil Engineers (s 31(1)). Emergency works can be carried out without submitting plans and sections, but a plan and sections must nevertheless be submitted to the authority for approval as soon as practicable after the work has been carried out (s 3(3)).

The undertakers' works must be carried out to the reasonable satisfaction of authorities affected and the street or controlled land must be restored promptly to its previous condition; subsidence or deterioration during the six months following the restoration must be made good (s 7). The liability continues beyond this period, however, if the reinstatement is negligently carried out (*Hartley* v *Rochdale Corporation* [1908] 2 KB 594). In that case an accident was caused by negligent reinstatement of a pavement surrounding a waterbox which had been repaired and reinstated. The liability under the Waterworks Clauses Act 1847, s 32, was for three months after the work was done. The accident happened two years and nine months after the work was done, but the corporation (which reinstated as agent for the water company) was liable to the plaintiff, who had been injured in stepping off the box.

Where undertakers in carrying out code regulated works cause damage to any street or bridge, they are liable to the street or bridge authority, s 18(1). Where a water main which had been lawfully laid under a street burst and caused damage to the street the water authority was not liable under s 18 unless there had been negligence on the part of the water authority: *Department of Transport* v *North West Water Authority* [1983] 3 WLR 707 (HL). Section 18(2) relates to 'gas, electricity, water or any other thing required for the purposes of a supply or service afforded by any undertakers'.

The legal position where undertakers cause damage to telegraphic lines of public telecommunications operators is discussed in Chapter 17.

The street authority or managers may themselves elect to carry out the work of reinstating a street or controlled land which has been broken up or opened (except for the minor works, such as on service pipes or lines, previously listed: Sched 3, para 1(1)). They must give notice of their election to the undertaker, and the work is done at the reasonable cost of the undertaker (Sched 3, para 4(3)).

Similar provisions apply where undertakers carry out works under the code which involves breaking up or opening a public sewer. They are under a duty to reinstate, and in default of their doing so the sewer authority (the water authority: Water Act 1973, s 14) may reinstate; the sewer authority may also elect to reinstate without their being in default (s 12(3), (4) Public Utilities Street Works Act 1950). The sewer authority may also elect to do the part of the code regulated works itself which involves breaking up or opening the sewer (s 12(4)). It may also claim to have done by the undertakers any other works rendered necessary by the code-regulated works to make proper provision for drainage for which a public sewer of the authority is used at the time of the works (s 12(1)). So if the works will interfere with a public sewer the sewer authority can require to be done any additional works which will prevent interruption of the drainage. The sewer authority can elect to provide these additional works itself at the undertakers' cost (s 12(4)). If any code-regulated works include the breaking up or opening of a private sewer, drain or tunnel the undertakers must reinstate and make it good (s 13(1)). If they fail to do so the body or person having the management of the sewer, drain or tunnel may reinstate and recover the cost from the undertakers (ibid).

The 1950 Act also deals with the situation where apparatus is affected by work carried out to roads and bridges. Undertakers (called 'operating undertakers') proposing to carry out in a street or on controlled land any works under a statutory power or authorisation which are likely to affect protected apparatus of an undertaker owning it may not begin to do the work without giving notice of intention to the owning undertaker (s 26(2)). 'Protected apparatus' is any undertakers' apparatus (except that of a railway undertaking) in a street or controlled land maintainable under a statutory power or authority (s 26(1)). Three clear days' notice must be given and the notice must state the nature and place of the proposed works (s 26(2)). Notice is not required for work on a service pipe, service line or overhead telegraphic line or in a case of emergency (but in this last case notice must be given as soon as reasonably practicable after the work has been done) (ibid). The owning undertakers are entitled to have reasonable facilities for supervising the work (s 26(3)). If damage is caused to the apparatus by the operating undertakers they must compensate the others to an amount equal to the expense reasonably incurred in making good the damage (s 26(6)), unless the damage occurred through the negligence or misconduct of the owning undertakers or their contractors (ibid). Where there is to be tunnelling or boring under protected apparatus it must be properly supported, and where an electric line is to pass near protected apparatus the line must be properly insulated and must not touch or conduct current through the protected apparatus (s 26(4)).

See Chapter 10/2(*b*) as to s 156 of the Highways Act 1980, which attempts to control the successive breaking up of the highway by different undertakers.

Ministerial Circulars concerning street works can be found in the *Encyclopaedia of Highway Law* (Sweet & Maxwell) Vol. 3, pt VI.

5 Liability of highway authorities

By s 41 of the Highways Act 1980, the authority which is for the time being the highway authority for a highway maintainable at the public expense is under a duty to maintain the highway. By s 51(1) a person who alleges that a way or bridge maintainable at the public expense is out of repair may serve notice on the highway authority and after one month apply to the Crown Court for an order requiring the authority to put it into repair (subs (2)). By s 58(1) it is a defence for the highway authority to establish that it had taken such care as in all the circumstances was reasonably required to secure that the part of the highway to which the action relates was not dangerous for traffic. By subs (2) relevant factors include the character of the highway and the traffic which was reasonably expected to use it; the knowledge of the authority as to the likelihood of danger; and the standard of maintenance appropriate for a highway of that nature.

Chapter 16

Pipe-lines

1 Pipe-lines Act 1962

The purposes of the Act are to place the control of cross-country pipe-lines in the Secretary of State, to be exercised in the public interest; to enable those wishing to construct pipe-lines to do so without a private Act; to enable the Secretary of State to regulate safely the construction, operation and maintenance of pipe-lines; and to make the pipe-lines rateable.

1.1 Definition and scope

Section 65 defines a pipe-line within the Act as a pipe or system of pipes (with any apparatus and associated works) for the conveyance of anything other than air, water, water vapour or steam. The term does not include a drain or sewer, pipes for heating, cooling or domestic purposes, pipes used in building or engineering operations, pipes within the boundaries of an agricultural unit and designed for use for agriculture (defined in s 66), pipes in premises used for education or research, or a pneumatic dispatch-tube. Pipes of a public gas supplier, the electricity boards, the Central Electricity Generating Board and the United Kingdom Atomic Energy Authority are excluded (s 58). Much of the Act does not apply to pipe-lines of railway undertakers, dock pipe-lines or government pipe-lines (ss 59, 61, 62), or to pipe-lines in factories, mines, quarry premises or petroleum depots (s 60). Cement and chalk slurry, oxygen, brine and ethylene gas are now being piped. There is an oil pipe-line 254 miles long linking the Thames and Mersey estuaries ((1969) *The Times*, 20 March), and a natural gas pipe-line running from Italy to the USSR ((1978) *The Times*, 23 March).

1.2 Control of construction

Sections 1 and 2 say that a cross country pipe-line (one more than ten miles long: s 66) may not be constructed without the consent of the Secretary of State by means of a pipe-line construction authorisation, and that a local pipe-line (one not exceeding ten miles in length: s 66) may not be constructed unless a notice of intention to construct has been served on the Secretary of State not less than sixteen weeks before construction begins.

Part I of Sched 1 sets out the particulars to be given on application for pipe-line construction authorisation, such as three copies of a map of the proposed route, particulars of any consents needed and obtained and a statement of what will be conveyed in the pipe-line (para 1). Notice of the application to the Secretary of State must be published in the *London Gazette* and be given in any other manner that he may direct, and must indicate where the map can be inspected and how objections can be lodged. Notice must also be served on the local planning authorities through whose area the pipe will go and on any other person that the Secretary of State may specify (para 3). He must hold a public enquiry if a local planning authority objects to the proposals; in the case of any other objector he may either hold a public enquiry or allow the objector to appear before a person appointed by him for the purpose; he may hold a public enquiry even though no local planning authority has objected (para 4).

If the Secretary of State grants an authorisation, construction work must be substantially begun within twelve months of the grant, but he may grant an extension of time (s 1(4)).

By the Pipe-lines (Limits of Deviation) Regulations 1962 (SI No 2845) a deviation of not more than one hundred feet laterally from the route shown on the deposited map is permitted.

Where notice is given to the Secretary of State of intention to construct a local pipe-line the notice must give similar particulars to those listed in the case of a cross-country pipe-line (and the applicant must send a copy of the plan to the local authority: s 35(2)), and again the work, if permitted, must be substantially begun within twelve months of his consent (s 2). There is no provision for advertisement, objection or public enquiry in this case, except that the Secretary of State may by statutory instrument direct that the procedure of s 1 shall apply to applications for the construction of local pipe-lines of the class specified in the order (s 6). By a new s 6A (added by s 25 of the Petroleum Act 1987) a pipe-line construction authorisation shall not allow a modified route for a pipe unless at least 28 days notice has been given to the local planning authority and any person specified by the Minister. If there are objections a public enquiry must be held. If an addition to a pipe-line already less than ten miles long would make it exceed ten miles, or the joining of two or more would make them in total exceed ten miles in length, then the procedure of s 1 for cross country pipe-lines must be followed (s 7).

The procedure for an application to lay must be followed where there is a wish to divert a cross-country or local pipe-line (s 3). Where the Secretary of State grants authority to lay or divert, he may at the same time grant the appropriate planning permission (s 5(1)). The execution of works for inspecting, maintaining, adjusting, repairing, altering or renewing a pipe-line (including breaking open any street or other land) does not constitute 'development' and so does not require planning permission (s 5(2)). Emergency works of construction or diversion may be done without the Secretary of State's prior permission provided permission is obtained as soon as reasonably practicable (s 8).

A person to whom a pipe-line construction or diversion authorisation is granted must forthwith after the grant deposit with each local authority through whose area the pipe is to run a copy of so much of the plan submitted with the application to the Secretary of State as shows the part of the route that lies within that area (s 35(1)). These plans must be open to inspection free of charge at all reasonable hours at the local authority's offices (s 35(5)).

1.3 Compulsory acquisition

The Secretary of State may authorise the compulsory purchase of land by a person proposing to lay a pipe-line (s 11(1)), or may authorise the placing of a pipe-line in land by means of a compulsory rights order (s 12(1)). Where compulsory purchase is applied for to the Secretary of State, the applicant must follow the procedure of Sched 2, Pt I. The applicant lodges with the Secretary of State three copies of a large scale map (not less than six inches to the mile), showing the land proposed to be acquired (para 1 (*b*)). The Secretary of State may either reject the application or order that it be allowed to proceed (para 2). If he allows it to proceed there must be advertisements on two successive weeks in one or more local newspapers stating that the application has been made, and saying where the map has been lodged and where objection can be made: a notice of the application must also be served on every owner, lessee and occupier (except tenants for a month or less) giving at least twenty-eight days in which to object to the Secretary of State (para 3). The form of the notice is prescribed in the Pipe-lines (Notices) Regulations 1963 (SI No 151). Where an owner, lessee or occupier objects the Secretary of State may either hold a public enquiry or allow the objector and applicant an opportunity of appearing before a person appointed by him (para 4). The procedure is governed by the Pipe-lines (Inquiries Procedure) Rules 1967 (SI No 1769), operative from 1 January 1968. Where the Secretary of State refuses to grant a compulsory purchase order he must give the applicant a written statement of his reasons for refusing (para 5). Where the order is granted the applicant must publish the fact in one or more local newspapers, describing the land and saying where a copy order and map may be inspected, and must serve a similar notice on each owner, lessee and occupier of land comprised in the order (para 7(2)). Appeal to the High Court within six weeks of the order becoming operative is possible on the grounds only that the order made was not authorised by the Act or that any requirement or regulation of the Act has not been complied with (para 9(1)).

The same procedure, with the appropriate modifications set out in Sched 2, Pt II, applies to applications for compulsory rights orders. Where the Secretary of State grants such an order he may attach such conditions as he thinks fit concerning (inter alia) the manner, method or timing of the carrying out of the works authorised (s 13(1)). The Fourth Schedule sets out the ancillary rights which may be conferred by a compulsory rights order, such as a right of access to the pipe by the pipe owner, with or without vehicles, and to deposit on the land temporarily materials brought by such vehicles and required in connection with the pipe; a right to place and renew markers to indicate its position; a right to erect stiles, gates, bridges or culverts to facilitate access to the pipe; and a right to construct works named in the order to facilitate maintenance, inspection or protection from damage.

1.4 Compensation

By the Land Compensation Act 1965, where an order for compulsory acquisition is made a person cannot be made to sell part only of a house, building or manufactory, or of a park or garden belonging to a house, if he is willing to sell the whole, unless the part of the premises can be taken without material detriment to the rest, or the part of the land can be taken without seriously affecting the amenity or convenience of the house (s 8); if a part only is taken, compensation shall be payable for loss caused by severance (s 7). Account shall not be taken of works which in the view of the Lands Tribunal were done on the land with a view to obtaining compensation or increased compensation (1962 Act, Sched 3, para 3).

Where an order for compulsory rights acquisition is made and a person entitled to an interest in the land affected proves that the value of his interest is depreciated by reason of the making of the order, he shall be entitled to compensation equal to the amount of the depreciation (1962 Act, s 14(1)). There is also a right to compensation for damage to, or disturbance in the enjoyment of, any land or chattels by reason of the exercise of any rights conferred by a compulsory rights order (s 14(2)). An example would be damage to crops caused by the pipe-line owner exercising his right of access to the pipe for inspection or maintenance conferred in Sched 4.

1.5 Pipe-lines in streets

Pipe-lines may be laid in a street which is a maintainable highway with the consent of the highway authority, in a street which is prospectively a maintainable highway with the consent of the highway authority and the street managers, and in any other street with the consent of the street managers (s 15(1), (10)). The pipe-line owner may open and break up the street to lay, inspect, maintain, adjust, repair, alter, renew or change the position of a pipe-line (s 15(1)). The consent of the appropriate authority may be given on reasonable conditions but must not be unreasonably withheld (s 15(3), (4)). Where the pipe-line is to be placed along a line crossing the street, consent may not be refused unless there are special reasons (s 15(2)). The Secretary of State will determine any disputes as to consents and conditions on the placing in a street of a pipe-line (s 15(6)). Similar disputes as to the placing of a pipe-line along a line crossing the street go to an arbitrator (s 15(7)). The codes of the Public Utilities Street Works Act 1950 (see Chapter 10/4) apply where pipe-lines are to be laid in streets, as modified by ss 16 and 17.

1.6 Safety of pipe-lines

The Health and Safety Executive may by notice to the pipe-line constructor specify in respect of a pipe-line or any part of it the manner of construction, the materials and components to be used and the depth at which the pipe is to be (s 20). A pipe-line owner may be prohibited from using any pipe-line which does not comply with the notice (s 22). The Health and Safety Executive may at any time by notice served on the owner impose such requirements as the Executive thinks fit in the interests of safety for the examination, repair, maintenance, adjustment and testing of the line and the inspection of the route of the line (s 23(1)). The Secretary of State may order, by notice, the owner to carry out works specified in the notice to make an abandoned or disused pipe-line safe (s 25). A pipe-line may not be used for the convey-ance of a different class of thing from that originally authorised without the owner giving at least three weeks' notice of intention to the Health and Safety Executive before making the change (s 26).

With the object of preventing free access to a pipe-line from being obstructed, ss 27 and 31 enact that, if a person erects a building or structure so that any part of it is less than ten feet from a point on the surface vertically above the pipe, or if a person deposits any earth, refuse, spoil or other materials within the same distance, the Secretary of State may by notice order him to demolish the building or carry out other work to protect the pipe, or, in the case of the deposit of earth, etc, he may remove it at the expense of the person who deposited it. The Secretary of State may serve on the owner of the building notice of the time (not being less than 21 days from the date of service of the notice) and place at which the question of ordering demolition will be considered and the owners of the building and the pipe-line are entitled to be heard when the question is considered (s 27(2)).

1.7 Undertakings to safeguard landowners

A series of safeguards has been agreed by the Country Land-owners' Association and the National Farmers' Union with certain pipeline promoters. The safeguards are listed in a publication of the Country Landowners' Association entitled 'Wires, Pipes, Pylons' obtainable from the Association, 16 Belgrave Square, London, SW1. The safeguards cover such items as the depth of the pipe-line (not less than three feet underground), trial borings, notice of entry (not less than one month), restoration of the land and hedges, ditches, land drains and watercourses, avoidance of pollution of water supplies or interference with water supplies and drains, insurance of the pipe-line by the company laying it, width of working strip during construction (normally 40 feet only), cathodic protection of buildings, compensation for loss of future development rights, protection of fishing and sporting rights, erection by the company of stock proof fences to keep out stock whilst the pipe is laid, notice of intended inspection and carrying out of works (the right to inspect and carry out works to be suspended if the area is infected because of foot and mouth disease, fowl pest, swine fever or other notifiable disease), and provisions for rendering the pipe-line and ancillary works harmless on abandonment.

2 Government oil pipe-lines

2.1 Permanent power to maintain

The Requisitioned Land and War Works Act 1948 ('1948 Act') made permanent certain government war works which were originally authorised only temporarily. Any government oil pipe-line may continue to be used provided that the original rights to maintain and use have been registered as local land charges or endorsed on the title deeds of the land affected by it (1948 Act, s 14). A government oil pipe-line is defined in s 12 as any government war works being the whole or part of a main or pipe installed for the transmission of petroleum. Where there is a right to retain and use a pipe-line, there is ancillary to it the right to remove or replace the pipe-line and by agreement to divert it (ibid, s 12). Where an oil pipe-line or accessory works are to be abandoned, notice of the intention to abandon must be given by the Secretary of State for the Environment to every owner and occupier (ibid, s 15). The Secretary of State must thereupon remove all sections of the pipe above ground and restore the land to its former level unless otherwise agreed with the owner and occupier (ibid), or unless in the Secretary of State's view it is not reasonably practicable to do this (Government Oil Pipe-Lines Regulations 1959 (SI No 715), reg 4). The pipe-lines and other works if not abandoned must be kept in good repair and persons who are caused injury or loss by a failure to repair must be compensated (1948 Act, s 15).

2.2 Compulsory acquisition

The Secretary of State may acquire any land required for constructing oil installations essential to defence; any land containing oil installations which were government war works immediately before 7 July 1948; and any easement necessary for the full enjoyment of land on which an installation is or is to be made (Land Powers (Defence) Act 1958, s 13) ('1958 Act'). Where the land cannot be obtained by agreement the Secretary of State has compulsory powers and he can take possession of the land immediately after giving notice to treat to the owner and occupier, subject to giving not less than fourteen days' notice and to paying compensation (1958 Act, Sched 2).

2.3 Wayleave orders

Oil pipe-lines may be laid under wayleave orders made by certain ministries where the pipe-line is needed for defence purposes (1958 Act, s 14). The right to lay comprises also the right to maintain, replace and remove (ibid, s 25). The order is registrable as a local land charge and will not bind a purchaser of a legal estate for money or money's worth in land affected if it is not registered (ibid, s 17). Compensation is payable for damage or depreciation to the value of land caused by the laying of the pipe-line (ibid, s 18). The ministry making the wayleave order may be ordered by Treasury regulations to keep the pipe-lines in good repair, and indemnify for loss or damage caused if it fails to do so (ibid, s 15). Regulations have been made under ss 15 and 18 as to repair and compensation for depreciation in the value of land because of the wayleave order: Government Oil Pipe-Lines (No 2) Regulations 1959 (SI No 724).

2.4 Protection

Where an oil pipe-line or accessory works are made or proposed under the various powers outlined above, the construction of buildings, roads or hard standing for vehicles and the excavation or deposit of earth, refuse or spoil may not take place within ten feet of the surface of the land immediately above the pipe-line or works without the consent of the Secretary of State (1958 Act, s 16). Where a wayleave order is made, mining under or near the pipe-line can be prevented subject to the payment of compensation for the minerals forbidden to be mined (ibid, s 15).

3 Petroleum and submarine pipe-lines

The Petroleum and Submarine Pipe-Lines Act 1975 set up a body corporate, the British National Oil Corporation (s 1), and gave it the power (among other things) to search for and get petroleum existing in its natural condition in strata in any part of the world, to move, store and treat petroleum and anything derived from it and to buy, sell and otherwise deal in petroleum and anything derived from it (s 2(1)). These powers included a power to provide and operate pipe-lines, tanker ships and refineries in connection with petroleum (s 2(2)(*a*)). By the Oil and Pipelines Act 1985 the British National Oil Corporation was dissolved and its property rights and liabilities were vested in the Oil and Pipelines Agency (ss 1, 2, and 3). The 1985 Act repealed ss 1–19 inclusive of the 1975 Act.

By s 20(1) of the 1975 Act a pipe-line may not be constructed or used in, over or under any 'controlled waters' without an authorisation in writing from the Secretary of State. 'Controlled waters' means the territorial sea adjacent to the United Kingdom and the sea in any designated area within the meaning of the Continental Shelf Act 1964, and 'controlled pipe-line' means any pipe-line in, under or over controlled waters (s 20(2)). 'Pipe-line' means in Pt III of the Act (ss 20 to 33) a pipe or system of pipes (excluding a drain or sewer) for the conveyance of anything, together with any apparatus and works associated with such a pipe or system. The definition was widened by s 25 of the Oil and Gas (Enterprise) Act 1982 to include 'any apparatus for treating or cooling any thing which is to flow through, or through part of, the pipe or system'.

A works authorisation for a controlled pipe-line may contain such terms as the Secretary of State thinks appropriate, including terms as to duration, route, things authorised to be conveyed, the persons who may be permitted to acquire an interest in the pipe-line, the methods of operation, and information to be furnished in respect of it (1975 Act, s 21(3)). Only a body corporate may be issued with an authorisation (s 21(2)).

By Sched 4, para 1, regulations are to be issued to provide for applying for a works authorisation. On an application for a works authorisation being made to him the Secretary of State may either reject it and give reasons or give notice to the applicant that it is to be considered further (Sched 4, para 2). In the latter case the Secretary of State will give to the applicant directions for advertising a notice of the application, which will allow objections to be made to the application within 28 days of the advertisement or such longer period as the Secretary of State prescribes (para 3). A map of the route will be made available for inspection (ibid). The Secretary of State may direct the serving of a copy of the advertised notice on such persons as he specifies (ibid). Where the Secretary of State forms the opinion that the route proposed for the pipe-line or part of it ought to be altered in a particular manner for any purpose mentioned in para 6 (so as to enable others to use it, for example: para 6 (*b*)) he must serve notice of his opinion on the applicant and on anyone else he considers likely to be affected by the alteration (para 4). The applicant may then claim a hearing by the Secretary of State, to influence his opinion, and others may make representations in writing and may, at the Secretary of State's discretion, be allowed a hearing (para 5). When the Secretary of State has come to a decision not to grant a works authorisation he will serve on the applicant, and on any person on whom a copy of the notice of application was served, a notice of the decision, giving reasons (except for those which in his opinion it would be against the national interest to state) (para 8). If he decides to grant an authorisation he will serve a notice of the decision on the applicant and on persons who made representations following a notice of application; and publish a copy of the notice in the London, Edinburgh and Belfast Gazettes (ibid).

It will be noted that there seems to be no provision for the holding of a public enquiry into applications for a works authorisation.

When a controlled pipe-line has been constructed and built the Secretary of State can (subject to a right of objection of the owner of the pipe-line) require modification of the pipe-line to increase its capacity on the application of a non-owner (s 22), and he can require an owner to allow others to use the pipe-line subject to their paying for doing so (s 23). The Secretary of State may serve notice to terminate a works authorisation if the works have not been begun within three years of the authorisation (s 24(2)). If an authorisation ceases, the controlled pipe-line will vest in the Secretary of State (or the Crown Estate or Duchy of Cornwall or Lancaster in appropriate cases) (s 25(1)). By agreement it may be transferred from the Secretary of State and vest in a person who obtains a subsequent authorisation (s 25(2)).

The Secretary of State may make regulations for securing the proper construction and safe operation of pipe-lines, for preventing damage to pipe-lines and for securing the safety, health and welfare of persons engaged on pipe-line works (s 26; Submarine Pipe-lines Safety Regulations 1982 (SI No 1513) as amended by 1988 (SI No 1985); Diving Operations at Work Regulations 1981 (SI No 399)). He may appoint inspectors with powers to inspect pipe-lines and premises, vessels and installations used or intended to be used in connection with a pipe-line or proposed pipe-line (s 27; Submarine Pipe-lines (Inspectors etc) Regulations 1977 (SI No 835), as amended by Submarine Pipe-lines Safety Regulations 1982 (SI No 1513). By s 21 of the Petroleum Act 1987 there is an automatic safety zone around every installation which is established for (inter alios) the exploitation or exploration in or under the shore or territorial waters or waters designated under the Continental Shelf Act 1964, s 1(7) of mineral resources, or the storage of gas or the conveyance of things by pipes. This zone is 500 metres from any part of the installation (s 21(5)). A safety zone may also be established by Ministerial order around an installation in transit to its station (s 22). Under the Offshore Installations (Safety Zones) Regulations 1987 (SI No 1331), in force from 1 September 1987, the safety zone provision does not apply to a vessel entering the zone in connection with laying, inspection or repair of a submarine cable or pipe line.

Part I of the Petroleum Act 1987 contains measures for dealing with the abandonment of offshore installations.

Sections 34 to 39 of the 1975 Act, which dealt with the construction and extension of refineries under authorisation from the Secretary of State, ceased to have effect from 9 June 1987, by s 28 of the 1987 Act. An application to establish a crude oil refinery or an installation for the gasification and liquifaction of 500 tonnes or more of coal or bituminous shale per day cannot be considered for the grant of planning permission unless it is accompanied by an environmental impact assessment under the Town & Country Planning (Assessment of Environmental Effects) Regulations 1988, (SI No 1199) in force from 15 July 1988.

Chapter 17

Telecommunications and Data Processing

By the British Telecommunications Act 1981, s 1(1), ('the 1981 Act') a public corporation called British Telecommunications (called in the Act 'the Corporation') was set up. It took over from 'the appointed day' (1 October 1981) some functions previously discharged by the Post Office. By s 10(1) before the appointed day the Post Office was to take steps to separate from the remainder of its undertaking the part concerned with the provision of telecommunication and data processing services. By s 10(2), on the appointed day all that part was transferred to the Corporation and by s 11(1) the rights and liabilities transferred included all those which by virtue of the Telegraph Acts (from 1863 to 1916, s 11(3)) were enjoyed by and incumbent upon the Post Office.

By s 2 of the Telecommunications Act 1984 ('1984 Act') the exclusive privilege of running telecommunications systems which was conferred by s 12(1) of the 1981 Act was abolished from the appointed day. A public telecommunications operator ('PTO') may run a telecommunications system if it holds a licence to do so issued by the Secretary of State or the newly-established Director General of Telecommunications, s 7(1) 1984 Act, ('DGT'). The duties of the Secretary of State and the DGT include promoting the interests of consumers; promoting effective competition between persons engaged in commercial activities in connection with telecommunications in the United Kingdom; and promoting the provision of international transit services (s 3(2), 1984 Act). Other functions of the DGT include keeping under review telecommunications activities inside and outside the United Kingdom, to investigate complaints relating to telecommunications services and apparatus supplied in the United Kingdom, and to investigate conduct which may be detrimental to the interests of consumers of telecommunications services or apparatus (ss 47, 49 and 50).

British Telecommunications was dissolved from an appointed day (s 69) and its property vested in a company nominated by the Secretary of State, British Telecommunications plc ('British Telecom').

By s 4(1) of the 1984 Act 'telecommunications systems' are defined as 'systems for the conveyance through the agency of electric, magnetic, electro-magnetic, electro-chemical or electro-mechanical energy, of:

(a) speech, music and other sounds;
(b) visual images;
(c) signals serving for the impartation (whether as between persons and persons, things and things, or persons and things) of any matter otherwise than in the form of sound or visual images; and
(d) signals serving for the actuation or control of machinery or apparatus'.

By s 5 it is an offence to run a telecommunications system without a licence. By s 6(1) it is not an offence to send messages by heliograph, for example, or by any other means where 'the only agency involved is that of light' which can be 'perceived by the eye and without more'; nor is it an offence to run an apparatus on premises in single occupation or in a vehicle, vessel, aircraft or hovercraft or in two or more of those mechanically coupled together. By s 6(1), 1984 Act the privilege is not infringed by a broadcasting authority's transmissions.

By s 24(1) 1984 Act the Secretary of State may authorise a PTO to purchase compulsorily any land which is required by it for or in connection with the exercise of its powers or which is reasonably foreseeably required. By s 37(1) a person authorised in writing by the Secretary of State may enter upon and survey any land (if not covered by buildings or used as a garden or pleasure ground) in order to ascertain whether the land would be suitable for use for the PTO's business. Similar provisions apply to Scotland and Northern Ireland (ss 38, 39). By s 41(1) the Acquisition of Land Act 1981 applies to a compulsory purchase of land by a PTO as if the operator were a local authority. By s 40(1) for the purpose of the acquisition of land by agreement many of the provisions of Part I of the Compulsory Purchase Act 1965 apply.

By s 3(1) it is the duty of the Secretary of State and the DGT to secure that there are provided throughout the United Kingdom so far as reasonably practicable such telecommunication services as to satisfy all reasonable demands, including in particular emergency services, public call box services, directory information services, maritime services and services in rural areas.

1 The telecommunications code

By para 1, Sched 2 of the 1984 Act, 'telecommunications apparatus' includes any apparatus defined in s 4 above and any apparatus not so falling but designed or adapted for use in connection with the running of a telecommunications system and in particular:

(a) any line, that is to say, any wire, cable, tube, pipe or other similar thing (including its casing or coating) which is so designed or adapted: and
(b) any structure, pole or other thing in, on, by or from which any telecommunications apparatus is or may be installed, supported, carried or suspended;
and reference to the installation of telecommunications apparatus shall be construed accordingly.

The Secretary of State may authorise any licensed PTO to use the Code if the operator is running a public telecommunications system or one for the benefit of the public (s 10(2)).

Under Sched 2 the consent of the occupier of land is needed for an operator to do works on land in connection with the installation, adjustment, or alteration of telecommunications apparatus, or to keep it installed on the land, or to enter the land to inspect, para 2(1). The owner of the freehold is bound by the agreement if he did not confer the right himself or agree in writing to its being granted, para 2(2), for as long as the occupier who granted it (not being a person with a lease of less than one year) remains the occupier. After the end of the granting occupier's term the freeholder or reversioner is not bound. Thus the operator would be well advised to obtain the consent of those with large interests in the land before investing heavily in apparatus in it. The person with an interest in possession who has not consented to the works can require the operator to restore the land to its former condition. A procedure is set out in para 21 to be followed by the person who wishes removal but the operator can serve a counter-notice and on application to a court may be allowed to retain the apparatus on the land, on terms. If the conferring of rights by the occupier reduces the value of the owner's interest in the land because of the need for the owner in due course to use the procedure in para 21 to have the apparatus removed, the owner may have compensation for depreciation of his interest and the amount is determined by the Lands Tribunal under the Land Compensation Act 1961 in default of agreement, para 4(4), (5) and (6).

2 Rights and powers of an operator

Street works may be carried out to install telecommunications apparatus under, over, along or across any street, or to inspect, maintain and adjust it, or do works incidental to those activities such as breaking up or opening it, tunnelling or boring under it or breaking up or opening any sewer, drain or tunnel and the code in the Public Utilities Street Works Act 1950 applies, para 9(1). If the street is not a maintainable highway then agreement is required of the occupiers under para 2 or a court order under para 5 dispensing with agreement, para 9(2).

Where the operator has telecommunications apparatus installed on land, he has the right to install and keep installed lines which pass over other land in the vicinity and connect to that apparatus but they must not be less than three metres above the ground or within two metres of any building over which they pass, para 10(1). This paragraph does not authorise the installation on the land of apparatus used to support the line or any line which interferes with the carrying on of any business taking place on the land, para 10(2). The occupier may object under para 17.

By para 12 an operator has power to cross any relevant land with a line and to install the line and other telecommunications apparatus on, under or over that land and to execute works of installation and repair. The line need not take the direct or shortest route to cross the land but it must not deviate from that route by more than 400 metres, para 12(2). The operator must, except in an emergency, give the person with control of the land at least 28 days' notice of his intention to enter and do the works, para 12(4). In para 12 'relevant land' means land used wholly or mainly for a railway, canal or tramway, para 12(10).

Where any tree overhangs any street and in so doing interferes with the working of any telecommunications apparatus or will interfere with apparatus which is to be installed the operator may require the occupier of the land to lop it; the occupier may by notice object and the operator may obtain a court order for it to be lopped by the operator, para 19(5).

Where an undertaker (defined in para 23(10) as including a local authority, gas or electricity supplier or water supplier or disposer of sewage, and other bodies such as a railway, road transport or canal authority) proposes to undertake works which will or may involve a temporary or permanent alteration of any telecommunications apparatus on the undertaker's land notice of at least ten days must be given to the operator. The operator may serve a counter-notice saying that he intends to make the alteration proposed or that the alteration must be done under his supervision, para 23(4).

The Town and Country Planning General Development Order 1988, Sched 2, parts 23 and 24 gives permission for certain developments without the need for an express planning permission. For example, the installation of telecommunications apparatus (other than on a building or other structure) which does not exceed 15 metres in height; and in the case of apparatus on a building or other structure, an installation not exceeding the height of the highest part of the building by 10 metres in the case of a building 30 metres or more in height, is permitted, part 23. The installation of a microwave antenna is dealt with in part 24.

Circular 22/88 issued with the Town and Country Planning General Development Order 1988 says in respect of article 4 directions concerning telecommunications equipment,

> Telecommunications by code operators will normally be subject to a reasonable degree of control in the licences issued to individual operators under the Telecommunications Act 1984. Blanket article 4 directions to control Sched 2, Part 24 development will not normally be approved. However, different considerations may apply where urban or rural locations of special attractiveness seem likely to attract particularly obtrusive or inappropriate telecommunications developments.

3 Compensation

Where a right conferred by the Code is exercised over another's land compensation is payable by the operator under the Compulsory Purchase Act 1965, s 10, if the amount of the compensation exceeds £50, paras 16(1) and (3).

Terms for compensation for the laying of cables and the passing of wires over land have been agreed between British Telecom plc, the Country Landowners Association and the National Farmers' Union. The rates are those payable by the electricity industry and they are set out in CLA Subscriber's circular 8/81, as amended at 1 August 1983.

Chapter 18

The Channel Tunnel

The Channel Tunnel Act 1987 provides for the construction and operation of a railway tunnel system under the English Channel, with associated works. It provides (in the words of the preamble) for improvements in the road network in Kent which will serve it and in the rail network in South Eastern England and for the incorporation of the railway tunnel system into the rail system of the United Kingdom.

1 Purpose

The primary purpose of the Act is to provide for the construction and operation of a rail tunnel link under the English Channel between the United Kingdom and France in accordance with a treaty signed in Canterbury on 12 February 1986 and in accordance with that to improve the road and rail network in the area with a view to accommodating the traffic (s 1). Private Concessionaires are empowered to construct and maintain the works specified in Part I of Sched 1 to the Act, together with (in appropriate cases) Kent County Council and the British Railways Board (s 5). The land forming part of the tunnel system as far as the frontier shall together with so much of the surrounding subsoil as is necessary for the security of the tunnel be incorporated into England and form part of the district of Dover, Kent (s 10). The land comprising the seaward section of the tunnel system vests in the Secretary of State as it becomes occupied by the Concessionaires working from England, together with the necessary part of the surrounding subsoil (s 7).

2 Powers of acquisition

The Act authorises the Secretary of State to acquire compulsorily the land shown on the deposited plans (defined in s 49 as plans deposited in Parliament in April and July 1986) and other land to be acquired for authorised works by Kent County Council or the Railways Board (s 8). The Concessionaires, the Kent County Council and the British Railways Board may construct and maintain the works specified respectively in Parts I, II and III of Sched 1 (s 5). Planning permission is deemed to have been granted for the carrying out by the Concessionaires of 'such development as may be necessary or expedient' for carrying out the scheduled work, except to the extent that it consists of the erection of any hotel or building which is not required in connection with the movement through the tunnel of passengers or goods. There is also planning permission for the works of Kent County Council and the Railways Board (s 9). The Building Act 1984 Part I (building regulations) and the Building Regulations 1985 only apply in relation to a house or hotel as above and do not apply to the construction of the tunnel as a whole (s 31).

The scheduled works contained in Part I of Sched 1 include the construction of a railway and viaduct in Dover, roads in the district of Shepway and the borough of Ashford, railways in London running into Waterloo Station and associated rail works down to Ashford and Shepway.

3 Interference to existing works

By Part II of Sched 2 all reasonable precautions must be taken against interference with lines and electric wires and the Secretary of State may make regulations for regulating the use of electrical energy for the operation of the authorised railway (section A). The appropriate authority may use sewers or watercourses for the discharge of any water pumped or found during the construction of the scheduled works but first needs the consent of the appropriate water authority, internal drainage board or local authority (consent not to be unreasonably withheld) (section D). There are powers for the appropriate authority to interfere with the highway for the purpose of construction or maintenance of the works authorised by the Act and to construct bridges, all in consultation with the highway authority, (Part I of Sched 7). There are similar provisions for the protection of sewers (Part V, Sched 7), electricity undertakers (Part VI), drainage authorities (Part VII), and the Southern Water Authority (or its successor) (Part VIII), the Folkestone and District Water Company (Part IX) and telecommunications operators (Part X). The general procedure is that the appropriate tunnelling authority deposits plans giving particulars of the intended works to the particular body and agrees them with modifications to give effect to the reasonable requirements of the particular body; the work is done causing no more disturbance than is reasonable; the constructor makes reasonable compensation to the bodies for loss sustained and indemnifies them against claims made against them from the works; pays them any costs incurred by reason of the works (such as for altering other apparatus); restores the land as soon as practicable; and delivers to the bodies plans showing the position and level of the works carried out.

Index